THE PASTURES OF BEYOND

Also by Dayton O. Hyde

Fiction

One Summer in Montana
The Major, the Poacher, and the Wonderful One Trout River

Nonfiction

Sandy
Yamsi
The Last Free Man
Raising Waterfowl in Captivity
Don Coyote

For Children

The Brand of a Boy
Cranes in My Corral
Strange Companions
Island of the Loons
Thunder Down the Track
The Bells of Lake Superior
Mr. Beans

THE PASTURES OF BEYOND

AN OLD COWBOY LOOKS BACK AT THE OLD WEST

Dayton O. Hyde

Arcade Publishing • New York

Institute of Range and the American Mustang at the
Black Hills Wild Horse Sanctuary
Founded by Dayton O. Hyde in 1987
P.O. Box 998
Hot Springs, SD 57747
www.wildmustangs.com
Home where hundreds of wild mustangs live free.

Ways of Giving to help the horses: Gifts of Cash, Stocks, Real Estate, Bequests and Donations for Land Acquisition

Copyright © 2005, 2011 by Dayton O. Hyde

All Rights Reserved. No part of this book may be reproduced in any manner without the express written consent of the publisher, except in the case of brief excerpts in critical reviews or articles. All inquiries should be addressed to Arcade Publishing, 307 West 36th Street, 11th Floor, New York, NY 10018.

Arcade Publishing books may be purchased in bulk at special discounts for sales promotion, corporate gifts, fund-raising, or educational purposes. Special editions can also be created to specifications. For details, contact the Special Sales Department, Arcade Publishing, 307 West 36th Street, 11th Floor, New York, NY 10018 or arcade@skyhorsepublishing.com.

Arcade Publishing® is a registered trademark of Skyhorse Publishing, Inc.®, a Delaware corporation.

Chapter 22 appeared under the title "The Appy Mare" in *Thunder of the Mustangs*, produced by Tehabi Books and published by Sierra Club Books in 1997.

Visit our website at www.arcadepub.com.

10 9 8 7 6 5 4

Library of Congress Cataloging-in-Publication Data is available on file.

ISBN: 978-1-61145-328-7

Printed in the United States of America

This book is dedicated to
Slim Pickens, Mel Lambert,
Montie Montana, Dick Blue,
and every other cowboy that
forked a bronc or threw a loop.

THE PASTURES OF BEYOND

Prologue

THIS IS THE STORY OF A WEST THAT IS ALMOST GONE, a story of ranches, cattle, horses, Indians, cowhands, and rodeos, and of a kid from Michigan who ran away from home at thirteen to be a cowboy in Oregon. These days I need a stump to help get me on a horse, and the bad ones I rode get tougher with every telling. These are stories that you won't read anywhere else, and if they have a certain value, it is because I was there when they happened and even remember what too many have forgotten, how the West sounded and how it smelled.

Walk down any eastern Oregon street, and amongst a pollution of Californians you'll rub elbows with descendants of men I once knew, and you'll likely tread on the faded footsteps of men better than yourself because they lived at a time better than ours. Sometimes my memory fades, and I feel like a cowboy whose rope is too short and his horse too slow to catch the critter he's chasing. But there are times when I smell the heady odor of crushed sage under the hooves of a moving trail herd, and hear the snores of sleeping cowboys in their bedrolls, or the contented munching of tired horses, eyes-deep in a manger of timothy hay. My mind clears; the shadows on the mossy walls of my memory become vivid once more.

I'd best start by telling you about Homer Smith, because more than any cowboy I ever knew, he represents what I miss the most, leather-hided, strawberry-nosed, bull-voiced, ham-fisted, broad-shouldered, mule-skinning, yarn-spinning characters who made life hell or sometimes fun for those about them.

"God damn yuh, kid," he told me one day as he was about to die, "I like yuh!" After all the fights he'd won in barrooms, he was about to lose the big one. We hugged for the first time ever and turned away to hide our tears. Maybe I owe it to Homer to say a few words about him here, while I can see the twinkle in his sun-washed blue eyes, and see him standing bowlegged in the doorway of my memories. Here was a man who grew up in no man's mold, and God help the man even twice his size who called him a liar.

He'd been my uncle's foreman once on the old BarY, and came back years later, when I had bought the ranch from my uncle, to work for the kid he'd once tormented. He could lose tools faster than any cowboy I ever hired. Issue him a brand-new pair of fencing pliers or some side-cutters for cutting baling wire, and they would be missing before the end of the first day. Lost accidentally, or maybe chucked in meanness or frustration into the brush. The old cowboy always had an excuse. It was never his fault.

"Hadn't been fer thet damn old bay hoss, Tune, I'd still have them tools. Had 'em in my saddlebags when he bowed his old noggin and bucked me off fair and square. I spent all afternoon followin' his tracks back to the home corral. Could hev built you lots of fence hadn't been fer thet hoss. Them tools er bound to show up someday."

The old man was right. Show up they did, but it was fifty-three years later when I happened upon some of Homer's lost tools. I was checking fence in my pickup truck down on the Calimus Butte Ranch when I ran over a rusty hay hook and blew a tire. Climbing out of the vehicle in a blue funk, I kicked some rusty fencing pliers out of the dirt along with an old hammer. Long ago Homer and I had mended that piece of fence together. Both tools were rusted almost beyond recognition, and Homer, of course, was long in his grave.

I keep the pliers in a drawer in the ranch kitchen as a link to my past, a tool no one has wanted to borrow, handy for cracking nuts at Christmastime and not much else. But sometimes I take them out and sit before the woodstove, holding them in my hands just for memories of times back before World War II, when something terrible happened to the West I loved, and it was never the same again.

Partly it was Social Security that robbed the ranches of their old cowboys, enabling them to spend their final years in town instead of doing chores on the ranches they knew so well. Partly it was wartime industry with its big payrolls that lured them to cities and left them forever dissatisfied with ranch wages. Maybe some of them never went back to the farm after they saw Paree. Somehow, the old storytellers never came back to the old outfits, but frittered away their days in querulous company in some nursing home. Bereft of history and colorful tales by men whose vivid stories matched the turbulence of their lives, the ranches became pieces of real estate, nothing more.

We should have recorded the stories before those old cowboys took their last ride. So many adventures! So much of the history of the old, true West died with them and can never happen again. Ern Morgan, Fred Joy, Oliver Little, Buck Williams, Homer Smith, Al Shadley, Ernest and Etta Paddock, Mamie Farnsworth, Frank Emery, Orrie Summers, Buster Griffin, Tommy Jackson, Ash Morrow, Slim Pickens, Mel Lambert, all long dead, out riding the pastures of beyond. A list of names out of my personal past, men and women who lived lives, loved and were loved, drank whiskey like water, fought bare-fisted, and grubbed sagebrush by lantern light on land that, a half century or more later, is farmed by a stranger, unaware, uncaring even, of the richness that is history.

The traces are still there to find. But the story of the land lies fading fast in the piles of rusty cans and milk of magnesia and whiskey bottles in the dumps, in the tortured trees grown up between cast-iron spokes of old mower wheels, in bent drift pins which once firmed the logs of wild horse corrals, in the names of certain fields, or hills, or springs. In stories once told around a bunkhouse stove to kids who pretended not to listen. Without the stories that went with them, the heaps of old iron are just that, heaps of old iron.

We all have our memories, our own stories to tell, that stand only a feeble heartbeat from being lost forever, relegated to the dumps of old baby carriages and the rusting fenders of vintage cars that took us to church or witnessed the loss of our virginity. The stranger in the bar has his own list of folks who made up his personal history. I can only tell mine. As a writer I feel a responsibility to those whose lives touched my youth. Many of them were buried without an obituary, but these men and women deserve our thanks and our prayers. Important folk, for they shaped a West that was and will never be again.

These are the stories of old Indians I knew, old cowboys I knew, and old horses I knew. All gone now, traveling the great green ranges of heaven. This is partly my own story — of a lonesome kid, maybe the only one in history to run away from Marquette, Michigan, to ride saddle broncs and fight Brahma bulls in rodeos, who was lucky enough to grow up on one of the great cattle ranches of the West.

Book One

Chapter One

My life as a cowboy started amidst the thick upper foliage of a horse chestnut tree in Marquette, Michigan. I was a thirteen-year-old beanpole of a boy given to tree sitting and watching birds. I had been up that tree for several hours, watching a robin build a nest, when my full bladder told me I had better hurry home. But not wanting to disturb the robin, who seemed to be in the process of laying her first egg, I put the privacy of my treetop bower to good use and watched as a golden stream trickled groundward from leaf to leaf. As fate would have it, at that moment my mother was walking under the tree, heading up Spruce Street with her bridge group.

What really hurt my feelings was that of all the boys in the neighborhood, my mother knew exactly who was up that tree. The chorus of screams frightened the poor robin off the nest, and as the ladies rushed up the street to tend to their bonnets, I hit the ground running. Half an hour later I was hidden in a freight car just leaving town on a westward track.

I could have toughed it out, of course, and bared my behind to my father's leather slipper, but in my pocket was a crumpled letter. It was from my rancher uncle in far-off Oregon, who had written that he could step off the front porch of his ranch house and scoop up enough trout in a dishpan from the stream to feed his crew. If that were not enticement enough, his cowboys had just captured thirty wild horses and were breaking them to ride.

I felt no guilt at leaving home. For seventeen years my father had been bedridden with multiple sclerosis, and I had long been dreaming about relieving my parents' burdens by fending for myself. No matter now that I had neither spare clothes nor money, nor had I ever been more than sixty miles from home. I had yet to ride a horse, but at that moment I was already a cowboy.

By the time the steam locomotive had sounded its lonesome whistle at several crossings, however, I was so homesick and hungry that I was ready to jump off the first time the train slowed. My fate was decided by an old hobo named Gus, who, smelling of rancid sweat and cheap wine, threw his bedroll into my car at the next town, followed by a gunnysack that proved to be full of blackened pots, coffee, and dried beans. Nursing a hangover, he didn't talk much. For much of the trip I huddled near the open door, hypnotized by the rise and fall of telephone cables between the poles. Gus left me at a hobo camp under a trestle near Spokane, but not before he had lectured me on the dangers of rattlesnakes, booze, and young girls, and seen me safe from yard dicks on a Southern Pacific freight train headed south for Oregon.

When I jumped off the train a day later in the southern Oregon town of Chiloquin, I was tired, dirty, and as hungry as a teenager can be. Gus had given me fifty cents as a parting gift, and with this I purchased a chocolate cake from some ladies at a church sale, and sat down on the curbing to devour every crumb.

My uncle's ranch lay thirty miles to the east, one of the church ladies told me, but there was a stage that delivered groceries to the ranch and the logging camps beyond. I'd best find the driver, an Indian named Hi Robbins, and arrange to get a ride.

I found Hi Robbins at the back of the general store, loading groceries into the back of his Chevy panel wagon. The groceries were in different piles, one marked Pelican Bay Camp, and one marked Lamm's Camp. I took them both to be logging camps. The other pile, no less large, was marked for my uncle's ranch, Yamsi.

The vehicle was old and needed paint, but it was clean. I had never seen an Indian up close. I was astonished to note that Hi was as handsome and well-groomed as any western movie star.

At that moment, a slender Indian girl trotted up on a spirited black-and-white horse. She looked at me and giggled. I realized how dirty I must have looked to them both. I rubbed my face on the back of my torn sleeve, no doubt making matters worse, and tried to smooth down my cowlicks with my hand.

"Mr. Robbins," I said, "my uncle owns a ranch named Yamsi at the head of the Williamson River. Can I catch a ride out there in your truck?"

The man's bronzed face showed no emotion. "Inside," he said. "Inside the store there's a sink with soap and water. You better scrub the dirt and stink off yourself if you're goin' to ride with me!"

The girl must have thought better of me when I cleaned up. She slid from her horse and offered me the reins. "Go ahead," she said. "Ride him if you want."

"I got a sore leg," I said, not wanting her to know I was afraid. "Some other time."

"My name is Rose," she said shyly, as though a last name were superfluous. She seemed to sense that I was looking down her shirt front, and her face flushed darkly. Grabbing her reins and a mane-hold, she vaulted swiftly onto her horse and galloped off. I was ashamed at my own crudity and turned away, embarrassed.

Hi Robbins covered the groceries carefully with canvas and nodded to me to get into the truck. The roar of the engine filled the cab, making conversation impossible without shouting. We headed out of town on dusty roads that wound through great forests of ponderosa pines. Here and there were windfalls, so thick in the trunk a grown man could not have seen over them. Now and then we would pass small Indian ranches, where little brown children lined the fences to wave at Hi. Now and then older women in long skirts looked up from their chores, their round faces splitting in grins.

Once, we stopped at a roadside spring, where Hi sieved out pollywogs from a rusty coffee can with his fingers and filled the steaming radiator with water. "Your uncle know you're coming?" Hi asked. His voice was as refined and beautiful as his face.

I shrugged. "I guess not," I admitted.

"I figured as much," Hi said. "He's quite a character, that uncle of yours. Never did get married. They say he hates kids."

Doubts swept over me. My uncle had never really invited me west. Just told me about the trout in his stream and the wild horses. What if he got angry when he saw me and sent me packing back to Michigan?

We left the main road at a twenty-six-mile marker to drive across a cattle guard made of railroad iron and through a grove of virgin pines. There were no stumps anywhere, indicating the timber had never been thinned. The ranch house showed up suddenly while I was still ogling the immense trees. The building was huge and ominous, with sharply sloping roofs to accommodate heavy mountain snows. The substantial walls were made of local lava rock with windows framed by huge pine logs, showing rough scars where they had been peeled with a broad axe.

The windows caught the afternoon sun and stared back at us, each pane a tiny eye. I thought I saw a curtain move, but no dog barked and no one came out to greet us. Even the birds were silent, as though busy watching.

I helped Hi Robbins unload groceries onto the back stoop. "Good luck, kid," he said. The engine roared, leaving a pall of stinking blue oil smoke, and swiftly the old Chevy was gone up the dusty road. For a few minutes I could still hear the growl of his truck engine as he shifted gears, and see a cloud of volcanic pumice dust as he progressed up the ridge toward the logging camps to the east. Then all was silent, and I was alone.

There was water running everywhere amongst the willows of the house lot. It bubbled in clear springs from the ground and merged into a larger channel spanned by a crude log bridge. The holes in my worn sneakers squirted water as I crossed the bridge and moved across the wet meadow toward an A-frame barn and log corrals. There was a freshly skinned cowhide draped across a top log on the corral, raw side up, and a dozen black-and-white birds with long tails were picking it clean. I'd seen their pictures in my bird books. They were magpies and rattled off at my approach, cursing me angrily. From the rafters of the barn hung a fresh beef carcass, shrouded with cloth against the flies. The white cotton had soaked up blood, giving it the coloration of a strawberry roan horse.

Far off I heard a horse nicker and a cow bawl, but the sounds ceased before I could judge the direction. To my right was a small unpainted cottage I took to be a bunkhouse. The door of the adjoining outhouse swung open and shut with the wind; I peeked in, half hoping that there would be someone sitting there I could talk to, but the three seats were empty. On the floor of the privy was a pile of chewed pinecones where a squirrel had recently made a meal.

I was about to investigate the bunkhouse when suddenly I became aware that I was being followed. I turned and saw a tall gray man standing there, watching. "Who the hell are you," he growled, "and what are you doing nosing around?"

I felt a flash of resentment at the greeting. I sensed that this was my uncle, but I had expected him to be wearing cowboy boots, Levi's, and a big Stetson. Instead, wearing a wool cardigan sweater and a slouch hat, this man could be some old banker from Marquette.

"I was looking for my uncle," I smarted off. "He owns this place. You must be his hired man."

He watched me, unruffled, as though he hadn't heard.

"Deaf as a post," Hi Robbins had said. "He's got a hearing aid, but damned if he ever turns it on."

"Your ma was the pretty one in my family," my uncle said, as though he had read my lips and needed to put me in my place. "It would appear that none of her good looks rubbed off on you."

With that he turned back toward the house, hooking his head to indicate that I should follow.

The house had a musty smell of mildewed books and was dark and chilly. I had been around enough antiques in the big old houses in northern Michigan to know quality when I saw it, but there was a dark, somber cast of ancient walnut to the carved chests and highboys that depressed me.

He led me upstairs and down a long, gloomy hall to a north bedroom, where there was only a bureau, a small desk and chair, and a narrow cot. "You can sleep here," he said. "I've got business in town for a couple of days. There's food downstairs in the cool room."

I was looking at the cot, wondering how I'd fit, and when I turned he was gone and the hall was empty. I sat in my room, fighting homesickness, until I heard his car start and its sounds fade in the distance. Then I got out of that cold house to sit on the massive, sun-warmed lava rocks that formed a garden in front of the ranch house. Yellow-striped garter snakes slithered off to cracks in the rock as though they were not used to human intrusion. The shallow soil bore a profusion of wild roses, all bursting in pink as though to cheer me up. I sat until the sun had warmed me, and then began to explore.

The house was wired for electricity, but there were no power lines stretching from town. Instead, one of the outbuildings possessed a diesel generator that, I learned later, was only used during evening hours. Hanging on the kitchen wall was a crank telephone which rang every few minutes, jolting the silence with a variety of rings.

From the ranch, a single wire stretched from tree to tree through little white insulators. Listening in to conversations, I soon determined that the line ran to several ranches and various fire lookouts, with a central office at the Klamath Indian Agency, which handled emergency calls about forest fires. Penciled instructions on the kitchen wall informed me that in order to make a call one had to crank out two longs and a short to reach the agency operator, who would then complete your call. I had eavesdropped on some pretty interesting flirty conversations when someone heard my breathing and screamed at me to get the hell off the line.

That night, I lit a kerosene lantern with a big wooden kitchen match, cut two giant beefsteaks from a hindquarter hanging in the pantry or cooler, lit a fire in the big wood-burning cookstove, peeled some potatoes that Hi had brought from town, and cooked myself a meal fit for a thirteen-year-old. Then, having eaten to the point of discomfort and erased all evidence of my meal from the kitchen, I took the lantern and went upstairs to bed.

The next morning, I awoke shivering in the chill of the

fresh mountain air. A front had moved down off the Cascade Mountains to the west. Outside my window, pine boughs clashed like swords, parrying thrust with thrust. A flock of Canada geese, feathers ruffled, sailed past and sought shelter on the pond beside the house. A tree crashed nearby, and I could hear the scream of busting barbed wire and the snapping of wooden posts under the falling tonnage. Yet above the din, from the shelter of the gables, I could hear the persistent cooing of a mourning dove pleading for peace in the forest.

I was still apparently alone at the ranch, and my uncle would not be back until the morrow. I had ventured out into the dying storm to survey the damage when I saw a great cloud of pumice dust rising up like smoke through the pine trees south of the ranch and all but obscuring the sun. My immediate fear was of a forest fire, but soon I heard in the distance the bawling of cattle and the wild yells and curses of cowboys. First a trickle and then a deluge of thirsty cows rushed through the trees, found an open gate, and flowed toward the house springs to drink. Soon there were cowboys, faces caked white with dust, who eased through the milling cattle, dismounted, took the bridles off their horses so they could drink, then bellied down beside their mounts to thrust their own faces into the cold, crystal springwater.

The din hurt my ears. There were cows bawling for their calves and calves bawling for their mothers. A pair of white-faced Hereford bulls fought up a dust storm, battling at the edge of the willows, bellowing curses at each other, rattling horn on horn. Then the riders mounted again. "Move 'em out, boys!" someone shouted, and the yelling began once more. Reluctantly the herd left the water and knee-deep grass to move across the house lot into the corrals.

I stood on the bridge over the stream, watching from a distance as the cowboys rode through the corrals, mothering up calves with their cows. Soon the bawling stopped, and the cows, having nursed their calves, lay down to rest. In time, calves at their sides, the animals began to move out through the open gates and drift northward down a lane toward the home meadows along the Williamson River.

The cowboys unsaddled their mounts and turned them out to roll in the dust. The tired horses lay with their hooves kicking the air, then lurched to their feet and ambled out to pasture. Dark sweat marks showed on their backs and flanks. There was no running, kicking, or biting. It was the end of a long cattle drive, and their heads hung low with fatigue. From a hundred feet away, I could smell the sharp odor of wet horse hair and sweaty saddle blankets.

One by one and in pairs, the cowboys headed for the bunkhouse. A boy, not much older than I, nodded to me shyly. For a moment I thought he was going to come over and talk to me, but he changed his course swiftly and ran for the outhouse.

Black smoke poured from the bunkhouse chimney, then turned white as a fire took shape in the kitchen stove. Two men came out with buckets and scooped up water from the spring. I sat on the bridge and watched, too shy to go over and talk.

One at a time, men came out on the front stoop and emptied washbasins of pale gray water on the grass.

As I sat on the heavy planks, dangling my feet above the water, a big, white-headed horsefly bit me on the knee. I smacked it hard and sent the body floating down the stream, where a hungry trout made short work of it. My knee still hurt from the bite when I got up and headed for the ranch house to cook myself another pile of steak.

When he arrived from town the next day, my uncle seemed surprised to see the kitchen clean. He glanced into the cool room at the big chunk I had carved from the beef but made no comment. He seemed shy and unable to carry on a conversation with someone my age and was no friendlier than he had been when I arrived, but I gathered courage to ask him what had been uppermost on my mind since I arrived. I figured that he couldn't get more grouchy than he already was. "I was wondering," I asked, in a voice loud enough for even a dead man to hear. "I was wonderin', Uncle, when do I get to ride a horse?"

Chapter Two

I WASN'T SURE MY UNCLE HEARD ME ask about riding a horse, for he gave no sign, just turned and walked out of the house. I followed, but he ignored me, and so I wandered off. The bunkhouse and corrals were deserted, and I whiled away a few hours looking at wildflowers on the meadows and watching warblers mate in the willows, wishing that I had someone to talk to, wishing that I could learn to ride so that the next time Rose offered me her horse, I would be able to impress her.

I was shaking with hunger but wasn't about to ask the old man for food. That afternoon, just when my hunger pains were about to overcome my pride, the old man came bustling around a willow bush, tossed me a brown bag containing a meat sandwich he'd made for me, and told me to get into his car, that he needed someone to open gates.

I was to learn that my uncle hated to open gates and would put up with anyone handy as a passenger as long as he didn't have to get out of his car.

He drove a big, fancy, four-door Chrysler and treated it like a pickup truck, driving out over irrigated fields, rocks, and fallen trees as though it were impossible to get stuck. Whenever he drove down through the ranch, the crew kept a team of draft horses harnessed just in case. On this, my first ride with him, he had hardly gone a mile down the ranch when he gunned through a slough lickety-split, and buried that big green monster right to the axles. I started walking for help, but a couple of cowboys had been watching from afar and met me halfway with the team. The old cowboy driving the horses winked at me and shook his head. It was about the first friendly gesture I'd seen since my arrival, and my homesickness vanished in a flash.

While the men were hooking up chains to ease the car out of the mud, the old man took a shovel out of the trunk and informed me that it was time I learned how to irrigate. With the shovel over his shoulder, he stalked off across the flooded plain, inspecting the flow of icy springwater as it coursed down ditches and spread out over the grass through small cuts in the ditch bank.

Here and there he stopped to build a sod dam in a ditch, or put boards in a wooden irrigation box. I watched his every move, trying to learn. My legs grew numb from wading in cold water, and now and again I would leave a tennis shoe sticking in the mud and have to probe for it. Maybe deafness had affected his balance, for my uncle couldn't walk a straight line to save his soul from hell. Every time he changed directions unannounced, the shovel over his shoulder would come about like the boom on a sailboat and clout me alongside the head with a clang. Off he would go on another tangent.

The water flowed from a dam at the headwaters of the Williamson River, a crystalline stream rising from fault lines along the base of the pine-clad ridges. With that water, my uncle flood irrigated some two thousand acres of timothy, bluegrass, and clover, catching the water up in a series of small earthen dams, and redistributing it as it flowed down through the long valley.

Once the cowboys had used the team to pull the car to dry ground, the old man headed back, put his shovel in the trunk, and slammed the lid.

"Don't know how the hell I got stuck there," he muttered to the men. "Been through that slough a thousand times."

The men looked away so as not to be caught grinning.

A road of sorts wandered along the meadow at the edge of the valley. My uncle seemed to drive by memory, looking out over the meadows and seldom at the road. The branch of a pine tree smacked my window, causing me to jump and throw up my hands to protect my face.

Seeing me make a fool of myself put my uncle in a good mood.

"Those BarYs of mine are damn fine Hereford cattle," he bragged as we drove through the cows and calves that had arrived the day before. When the road finally skirted the dams and hugged the main channel of the Williamson, we had to stop at every bend to look down at the huge shadows of trout lurking like logs in the bottoms of the pools. I had already been smitten by Rose, the Indian girl; now I fell in love with the ranch. I was aching not only to ride a horse, but to get after those big trout.

During that trip down the valley, the man became more and more expansive, and I guessed that much of his early reticence was due to shyness. He drove with his hearing aid unplugged and the batteries eloquently placed in the ashtray where I could see them, indicating thus that I was not expected to interrupt with a comment. He never glanced at me when he talked, and might have been addressing the cows, the fish, or the pine trees.

I sat there as he drove, bouncing with every jolt, watching the scenery, thinking over what little I had heard about the man. He was my mother's favorite brother, and the son of the Reverend G. Mott Williams, bishop of the Episcopal diocese of northern Michigan.

Being the son of clergy, he grew up proving to other kids he was no sissy. He learned how to fight, and adopted the name of Buck instead of Dayton, maybe because it sounded tough. This I understood, for I had been named Dayton after my uncle but felt much more comfortable with the nickname of Hawk. To me the name fit right into my plans to be a cowboy.

At thirteen, Buck proceeded to make a small stake hunting deer for the logging camps, which set up a standard for my own independence. At fifteen he ran away from home to live with Cree Indians in the vicinity of Great Bear Lake in Canada. In one of the worst winters in history, he holed up in a small cabin and lived out the winter on beaver meat from his trapline. By spring he had developed scurvy, and as a result lost most of his hearing.

Across the ridge from him lived a Cree family, consisting of a man, his wife, several children, and an old grandmother. When the man froze to death on his trapline, the woman saved her children by feeding them the grandmother.

When spring came, the Royal Canadian Mounted Police took my uncle to Edmonton to testify at the woman's trial. Through my uncle's testimony, the woman was acquitted. "After all," my uncle delighted in saying, "they'd eaten the evidence."

According to my mother, when he tired of living a wilderness life he drifted south by railroad, through Portland, and then to San Francisco. His looks and charm brought him ready acceptance in Bay Area society, and he had enough income from his mother's inherited farms near Detroit to live rather well. A stage-door Johnny, he surrounded himself with actresses and lived on the top floor of the new St. Francis Hotel with his Airedale dog, captivating all who would listen with his tales of the far north.

Almost trapped into a social marriage, he sought the wilderness again, taking a job locating vast timber holdings in the Northwest for such railroad giants as the Harrimans, who were interested in investing some of their fortune in Oregon. While appraising timber for his new bosses, he stumbled upon that magic valley in southern Oregon that contained the headwaters of the Williamson River. In spite of a lack of major financing, he began putting together the large cattle ranching complex known as Yamsay Land and Cattle Company, whose brand was the BarY, and whose headquarters ranch was Yamsi. Both "Yamsi" and "Yamsay" were taken from Yamsay Mountain, high above the ranch, which Klamath Indians called "the home of the north wind."

My gentle mother lost faith in her adored brother when Buck visited his uncle Will Biddle in Portland, Oregon, and ran off with Will's wife, Margaret. On her divorce from Will, Margaret became my uncle's partner in his endeavors. That relationship was already fraying when I arrived at Yamsi, which may have accounted for the fact that the great stone house was often empty.

A year before the economic crash of 1929, Buck hired Margaret's son-in-law, a well-known Portland architect, Jamison Parker, to draw up plans for a large house at Yamsi, to be built out of native lava rock and huge ponderosa pines cut from the virgin forest around them. My uncle had a pal named Mortenson in Klamath Falls who owned a sawmill, and Buck was able to select the boards for the interior himself. These were hauled by steam locomotive over a logging track to a point ten miles from the ranch, and taken the rest of the way by team and wagon.

For a moment as I drove with my uncle, my reveries were shattered by a mule deer doe which darted in front of the car and plunged into the river. Lost in thought, my uncle didn't seem to notice that close call, and I went on with my thoughts.

The old man had a thing for towers, and the original plans included one as tall as the neighboring pine trees. Using cowboy help, Buck completed the house at a cost of eight thousand dollars. It included a foundation for the tower, but that edifice was never built, possibly because the Depression came along to put a brake on his spending.

Despite hard economic times, Buck and Margaret Biddle managed to turn the house into a showcase, filled with rare antiques from Europe and a superb collection of Oriental rugs. Such elegance tended to build a cultural gap between

Buck and the cowboys. He might own a fine herd of cattle, but the men would never accept him as I wanted to be accepted, as a real hand.

Being the boss's nephew and ignorant of all facets of ranch life, I was beginning to realize that I would have to battle hard to belong. No one had to tell me that to be respected by the men, I would have to beat them at their own cowboy games.

As I rode along in the big car, I reflected that I had been at the ranch three days and still hadn't ridden a horse or talked to a cowboy. With the impatience of youth, I felt that life was slipping past quickly and there was no more time to waste.

I was about to shout at my uncle and see if he had any vestige of hearing when he ran his Chrysler over a big pine stump. I flew up out of the seat, hit my head on the ceiling, and saw stars. My neck felt broken, and I moved my fingers carefully, happy to find that they still worked.

"Who the hell left that stump in the road?" Buck grumbled. He wasn't driving on the road, having left it a couple hundred yards back, but I wasn't about to make that point.

As he drove onward, he talked about places I'd never heard about, the BK Ranch, the Grigsby place, Wocus Bay, and Klamath Marsh. I looked at the instrument panel and saw that the radiator was overheating. I had no way of telling him, but pointed at the temperature gauge. He ignored me until a big cloud of steam arose from the radiator, obscuring the forest ahead. Fortunately we were still close to the river, and he let the engine cool, then carried up enough water from the river in his hat.

As we took off again, I hoped that my whole life here wouldn't consist of riding in his Chrysler, opening gates. The world I'd seen here so far was pretty small. Beyond the trout in the river and the horses I saw only at a distance, I was starting to think that maybe I should have stayed in Michigan.

What I did not appreciate was that at the time of my arrival on Buck's doorstep, the BarY was a great old-style livestock outfit, a rough, tough cattle and horse operation running seven thousand head of fine Hereford cattle and two hundred and fifty mostly Percheron broodmares. The land consisted of an eleven-thousand-acre holding on a vast plain at the foot of the Cascade Mountains, known as Klamath Marsh; a lease option on the BK, a great hay ranch in the Bly Valley; and the headquarters ranch, Yamsi, at the head of the Williamson River, plus ninety thousand acres or so of range leased from the Klamath Indians. Most of the cattle were shipped south every fall to winter on ranches in northern California, west of the town of Williams.

Despite its size, the total amount of rolling equipment on the BarY consisted of one old cabover truck with stock racks, one D2 Caterpillar tractor, a couple of pickup trucks, and my uncle's Chrysler. This equipment was rotated from one ranch to another according to dire need. The old man was tight with his money, and men and horses were cheap.

When cowboys moved from one ranch to another, they tied their belongings behind their saddles and took off a-horseback cross-country. There seldom were roads going where they wanted to go. The company was well known in the West less for its fine cattle than for other reasons. One frustrated cowboy described the BarY as being "just one big mismanaged emergency after another."

Buck still hadn't acknowledged my request to ride a horse, but the morning after our trip down the valley, I was rousted out of bed at three in the morning by the foreman, fed a plate of greasy fried eggs and bacon, stuck on a sleepy sorrel horse called Yellowstone, and forced to trot along with three silent cowboys to Klamath Marsh, some fifty miles away.

The old, worn Porter saddle issued me immediately began to shape my young bones to its contours, and the hair on the inside of my legs was worn off on the first mile and has never grown back. Every muscle of my body ached, but I would have died rather than complain. For three days we worked cattle on the Marsh ranch, separating out cows without calves as well as any that needed dehorning.

That job done, we headed east again, forded the Williamson at a place called Little Wocus Bay, and trotted the fifty miles or so back to Yamsi. Yellowstone seemed to me to be awfully rough-gaited, but he was my first horse and I had nothing to compare him to. The last mile was unbearable agony. I had a green-apple bellyache from trotting, and my eyes refused to focus. As I slid from the saddle, both knees buckled with excruciating pain, but I shoved myself off the ground and unsaddled Yellowstone by myself, then turned him out into the corral to roll with the other horses.

That night I soaked in a hot bath, carefully peeling off bunches of dead skin from my inner legs. By morning I was determined to ride again. At dawn, I scrambled myself some eggs and hurried over to the barn, hoping that the cowboys had not left without me.

The foreman, a tough little cowboy named Ern Morgan, seemed surprised to see me. Since Yellowstone needed a rest, he issued me a big gray horse named BK Heavy. I think that everyone expected Heavy to buck me off. The corral fence was lined with cowboys, but the big gray tolerated my clumsy efforts to mount, and soon I had joined two old cowboys who were headed thirty miles to the north, to gather any BarYs that had strayed off their range.

The two old men were more talkative than the others. As we rode north through the ranch, the cowboys told story after story, and I learned to keep abreast of them so as not to miss a word they said. Their talk was of horses and women. It took me quite a while to realize that the girls they discussed with evident affection were prostitutes in Klamath Falls. I could only guess at a female's anatomy, and there was nothing I could contribute to the conversation.

We found a few BarYs scattered amongst cattle belonging to the Kittredge outfit, and some Indian Department cows with conspicuous ID brands on their ribs issued by the Department of the Interior to an Indian, Charlie Lenz, in return for a share in the calf crop. I learned to read the ownership of cattle by their brands and earmarks, and by three that afternoon we had started driving the BarYs south toward my uncle's range.

At first the cowboys treated me like a chore they had been saddled with, but maybe they sensed my eagerness to learn, for they soon relaxed and even managed a grin or two at my blunders. One of the cowboys rolled a Bull Durham cigarette and handed it to me, but as I tried to light it, I sucked all the tobacco into my mouth and had such a fit of choking I scared my horse and almost landed on the ground. I grabbed the saddle horn with such a death grip that I was able to pull myself back on.

I glanced furtively at the cowboys, thinking they might be laughing at me, but their sunburned faces were cast in bronze. The cows at that moment chose to slip off into some jack-pine thickets, and we were busy for the next half hour trying to get them back on the trail.

Chapter Three

I HAD GOTTEN ALONG PRETTY WELL with Yellowstone and BK Heavy, considering that I was just a greenhorn or, in cowboy terms, a "button." With more confidence than ability, I begged Ern Morgan to let me ride a huge bay horse named Sleepy. I didn't realize that no one else wanted to ride Sleepy because the horse had a habit of groaning and grumbling as he walked, which got pretty old by the end of a long day.

I might have suspected by the way Morgan's lips puckered in a grin around his cigarette that the cowboys were in for some fun. I had no sooner hit the saddle than Sleepy groaned, grumbled, and ducked his head, bucking me off in one jump. I lit face-first in the dirt with my mouth wide open and came up spitting sand and whatever else was in the corral. The horse continued to groan and buck with big slow jumps.

He looked so easy from the ground, I could have kicked myself for not hanging in there and riding. Along the fence I could see cowboys laughing at me. When Sleepy finally stopped at the end of the corral, I limped over to him and got on. My knees were shaking with pure fear, and my foot kept jumping back out of the stirrup. The old horse still had a hump in his back, but I eased him off and followed two cowboys out the gate, heading south toward the Wildhorse Meadow country to look for strays.

It was an easy ride through open pine timber, and the cowboys kept at a walk. We had gone about four miles when Sleepy balked at crossing a small stream. I kicked him in the ribs, and suddenly the horse was bucking again and there was lots of daylight between me and the saddle.

"Lean back, kid! Lean back!" one of the cowboys called to me. I leaned back and finally caught the rhythm. After three jumps, Sleepy probably thought his luck had run out, for he settled down with a groan, trotted across the stream, and never tried to buck with me again.

Something happens in a kid's head when he manages to ride a bronc and not fall off. Down deep, I knew that old Sleepy was no world-beater, but what I learned was that I didn't have to fall off on the first jump. The next time a horse bucked with me, I knew that I could at least try to hang in there and ride.

With that one brief success in riding Sleepy, my mental attitude changed, as did my relationship with the other cowboys. The cowboy who hollered at me to lean back was Jack Morgan, the foreman's brother. He had two gold teeth in front that flashed as he grinned, and his bronzed face was arrogantly handsome. Jack was a natural teacher, and when I managed to stay aboard Sleepy, he seemed to think I qualified for a little help.

Suddenly, there he was riding alongside, showing me how to hold my reins and how to make my body flow with the horse. He even shook down his rope, made a loop, and showed me how to catch an imaginary calf, jerk my slack, and take my dallies, or turns, around the saddle horn. In those few hours of riding with Jack, I learned things I couldn't have picked up in a year in the saddle.

Jack advised me to think positive thoughts on a bucking horse and ride. "Gettin' bucked off hurts," he said, "an' most times it's a long way back to the home corral afoot."

Jack was riding a fine black colt named Spade, who was just making the transition from a braided rawhide hackamore around his nose to a spade bit in his mouth. In a little meadow, Jack put the black through his paces. The colt made long, sliding stops, and could set back on his hindquarters and spin on a dime.

"You ride with me, kid," Jack said, "an' I'll have you reining horses just like this one."

That evening, after I had unsaddled Sleepy and turned him out to graze, I got up enough nerve to wander over to the bunkhouse and sit on the porch with the men. I felt out of place until Jack started telling the crew about my ride on old Sleepy and how tight I had choked the saddle horn. I didn't mind being teased, for I was the center of attention. It was the general opinion of those cowboys that Sleepy couldn't buck off a wet saddle blanket. But I didn't know that at the time, and I was some proud.

On the bunkhouse porch was a hook-nosed, round-shouldered cowboy named Buster Griffin, who was generally accepted to be the best ladies' man and bronc rider in the lot. Talk got around to a BarY horse named Whingding, who just happened to have come over in a load of fresh horses from my uncle's hay ranch, the BK. Some of the cowboys admitted that they had met their match trying to ride that horse and would be willing to put a little money into the hat to see Buster try to ride him.

Whingding was hard to catch and left the herd to stand alone at the end of the corral. He was a stocky dark bay horse with a jagged lightning streak of a blaze down his face, a heavy mane that hung on both sides of his neck, little gimlet eyes, and a snort that sounded like the report of a buffalo gun. According to Jack Morgan, Whingding never bucked more than six inches off the ground, but hit so hard he could make blood run from a cowboy's ears.

While Buster was getting his rigging from the big A-frame barn, the rest of us ambled over to the corral to watch the fun. Jack Morgan roped Whingding around the neck with a backward horse loop he called a "hoolihan," and the big horse settled down and followed Jack over to the fence. There were a couple of white saddle marks on his back, indicating that the horse had had some hard rides. Other than squatting when the saddle dropped on his back, and blasting our eardrums with a snort, the animal acted like any other seasoned cowhorse.

Buster had managed to get one foot in the left stirrup when Whingding whirled away from him and dropped his head to buck. For two jumps Buster stood in one stirrup, and then managed to gain the saddle and find the off stirrup with his toe. Our war whoops only seemed to turn the horse on. He bucked across the corral, turned at the fence, and started back. We could see a little daylight now between Buster and his saddle. Suddenly, Whingding sucked back underneath himself and whirled, sending Buster flying headfirst into the heavy logs of the corral. There was a loud crack as a lodgepole pine log broke in two, showering the ground with splinters. For a moment Buster lay quiet, then wiped his bleeding nose on the back of his buckskin glove. He stared in fascination at the blood for a moment, then grinned sheepishly at the crowd.

"Sure is lucky," Jack Morgan chuckled. "Sure is lucky Buster hit that log with his head. Otherwise he might have hurt something."

"One big goose egg for Buster!" someone said, laughing, as we trailed back to the bunkhouse. The cowboys who had lost money on the ride didn't seem very happy.

I looked back over my shoulder at Whingding, who was standing quietly, waiting to be unsaddled. I was already daydreaming about taking a setting on that horse while Rose watched and making one helluva ride.

I had been at the ranch about a week when Ern Morgan told me to saddle Yellowstone and ride over to the willows in front of the ranch house to make sure there were no cattle grazing in the garden. Ern had a silly grin on his face that made me suspect I was being had, but I went anyway. I rode out of the willows, and there, sunning herself on the great gray rocks in front of the house, was a naked woman. Her long white hair was spread out on the rocks like clothing hung out to dry. Yellowstone snorted and almost lost me as she moved and sat up, looking at me. Calmly, she reached for her robe and pulled it on. Unruffled, she called to me as I tried to back my horse through the willows.

"Come back," she said. "I'm Margaret Biddle. Your uncle told me you'd be here."

She was a little China doll of a woman. When she stood up, her hair dangled below her waist. She had blue eyes that seemed to pierce my shyness and lack of experience. Her skin was wrinkled ivory, but there was a suppleness to her leap down from the rocks that made me misjudge her age by twenty years.

I could hear cowboys whooping and laughing over by the barn, and I wanted to run for it, but she made me tie Yellowstone to the fence and come in for lunch. It was to be my first of many awful meals. In the kitchen, a roaring fire burned in the stove and the heat blistered the painted wall behind. She opened a can of vegetable soup, slopped it into a pot, placed it on the cherry-red stove, then proceeded to forget it as she showed me how to make sandwiches. The butter and jam had to be spread just so, right to the edges of the bread, and by the time she had stood over me until I got it right, the soup had boiled dry and the kitchen was full of acrid smoke.

She added water in a cloud of steam and we ate the soup anyway, along with the sandwiches, which tasted no better than other sandwiches despite her lecture on how to make them. Clearly, she had had servants all her life or she would have starved.

I kept eyeing Yellowstone out the front window, wishing I could steal away and hide in the pine woods, but by now she considered me her personal servant, and after I had carried in a body ache of suitcases from her car, she issued me a little apron and made me sweep the house, hang curtains, mop the floors, then carry out rugs to beat for dust on the front rail fence. Yellowstone snorted at my activity, and I think the horse was laughing at me. It was dark before I finally slipped away and led him back to the corrals.

My uncle arrived late and wasn't at all glad to see Margaret. I felt better when I saw him take the batteries out of his earphones and lay them on the table for her to see. I was about to cut steaks for supper off the hindquarter of beef in the cooler when she came storming in and said that the meat hadn't hung long enough yet, and it was her policy to eat the cheap cuts first and save the steaks. The meat was already covered with green mold, and I wondered just how long she planned to wait before it was ready to cook.

She burned another can of soup and offered some to my uncle, but he said, "I'd rather starve!" and took off for town in his Chrysler. I hid out upstairs in my little bed until I heard her snoring in her corner bedroom, then crept downstairs and fried myself a big steak.

The next day, Buck sent a cook out from town. She was a big Irish lady with arms like tree trunks, who started immediately to make apple and cherry pies until a delicious fragrance filled the house. She gave me a piece, and, starved for someone to talk to, I regaled her with stories. I had consumed one whole pie when Mrs. Biddle came in from riding her horse and started treating the Irish lady like a servant. I could feel the tension building and stood powerless when the cook seized a butcher knife, ran Mrs. Biddle out of the room, and told her to stay the hell out of the kitchen. The cook stayed on another two hours until the old lady started talking to her through the dining room door, telling her what was wrong with the pies. That evening Ernie Morgan took the cook back to town, and I was treated to another meal of burned soup. But the pies I stole after bedtime were delicious.

Granted, Mrs. Biddle did have a hold on me. I was deathly afraid she would have me sent back to Michigan if I didn't humor her. When I put my first apron in the stove, she produced another frilly one and made me wear it. I hated dusting, but came to like doing windows because I could at least look out and see Morgan and the boys out riding the fields. Whenever there was a knock on the door I hid, since I didn't want any of the cowboys to catch me wearing that silly girl stuff.

To say something nice about the old lady, she was a great horsewoman. She had a beautiful Standardbred named Ginger, who was too snooty to mingle with the other horses and had to be wrangled separately. She would crowd Ginger up to a stump and pile on, riding straight in the saddle, and despite her seventy-some years, she put in enough miles on cattle drives to tire a young man. The cowboys would have preferred riding alone, but she was part owner, and try as they might, it was tough to start a ride without her.

The cowboys were bilingual. They had one language they spoke in her presence and another while chasing a wild cow out of the brush. They needn't have bothered. A year later, when I heard Margaret Biddle in an argument with my uncle, I knew why he took the batteries out of his earphones. She had a vocabulary that would have made the toughest cowhand blush.

I got to saddling my horse at dawn so she wouldn't put me to work, joining anyone leaving on an all-day ride. Some of the stories the cowboys told me as we rode were not for young ears, but I was so naive I didn't understand what they were talking about. "Years ago when I went to school in the town of Bly," Jack Morgan told me, "they had a big outhouse behind the school with one side for boys and the other for girls. It wasn't long before the boys pushed out a knothole, that they used to spy on the girls. My brother decided he'd do the other boys one better. Someone dared him to stick himself through the knothole, and he took the dare. Trouble was, there was a girl in there who grabbed him and went round and round that knothole. She held on while the other girls hollered for the teacher. Turned out he was the only one in our family that ever got circumcised."

For weeks I laid awake at night thinking about that story, wondering how the poor boy ever lived it down.

Chapter Four

I loved the early mornings at the ranch. Blended with the fragrance of pines outside my window, I could smell clover blossoms from the hay meadows in the valley and hear a myriad of songs as birds awoke joyfully from a night's slumber. In the distance cows bawled plaintively for their calves, and sandhill cranes shouted angrily at any who dared a flight over their nesting territory in the marshes. Wilson snipe winnowed as they strafed the valley with their booming cries and rose high again. Horned larks tinkled earthward, singing their faint music as they plunged. Horses nickered for absent friends, and hooves thundered loud on the plank bridge as the wrangle boy galloped a herd of horses toward the corral for the day's work.

Even in summer, the mornings had the nip of frost to them. I lay in bed and thought of Rose, wondering if I would ever see her again. In my daydreams I rescued her from charging bulls and stampeding horses. I loped with her across the lovely, flower-strewn Wildhorse Meadows, noticing the swing of her young breasts with the easy rhythm of her horse.

I saw her again sooner than I expected. It was suddenly haying season, and my uncle offered a haying job for any Indian who wanted work. The bunkhouse filled rapidly, and the overflow camped out in tents by the house spring. There were no hay balers in those days, and putting up loose hay took lots of labor.

In addition to the hay Buck's crew put up at Yamsi and the BK, the BarY bought stacks at various neighbors, and since loose hay couldn't be transported to the ranch, the cows had to be driven to the hay for wintering.

Buck had bought his neighbor Lee Hatcher's hay for years, and the only problem with the deal was that Lee was a friendly man who never could say no to a neighbor in need. He was always willing to lend out his hay machinery to friends, and by the time haying season came every year, Lee never had much equipment around to put up his own hay. This year, all Lee had left was an old truck no one wanted to borrow, two mismatched draft horses, and a dilapidated sulky rake. He was pretty much limited to putting up the hay by hand.

Since I was low man on the totem pole, I was sent down to help Lee put up his hay in his barn. The system we used was pretty basic. An old man named Joe Perry drove the team on the mower and rake, and when Joe had raked the cured hay into windrows, Lee would arrive with the truck, and we would fork the windrows onto the bed until we had a load. Lee would drive the truck to the barn, where his wife would help him unload it into the haymow.

Late that afternoon, as the truck came back for its fifteenth load, Joe Perry stopped his team, tied up the reins, and stepped off the rake to relieve his bladder. The truck made a swift U-turn and sped back toward the barn, empty. Minutes later, Lee appeared in the hay field, out of breath and angry.

"Joe," he roared, "that was my wife driving the truck. Why in hell did you have to pee right in front of her?"

"Aw, Lee," Joe Perry said. "I wouldn't get excited if I were you. After all, she only saw part of it."

That night I slept in the barn in the loose hay and was lulled to sleep by the contented snuffling of my horse, Yellowstone, and the work team as they ate new hay from the manger. I had blisters on my hands from the handle of my pitchfork, and my muscles ached when I moved, but I think I must have slept with a smile on my face. Each day that I lived in Oregon was bringing me a new adventure.

It was a week before I could get back to Yamsi. One evening I drifted down through the pine forest on Yellowstone and was relieved to see the big rock house standing in the trees. The horse seemed as glad as I was to be home. He nickered to friends in the corrals, as I looked around amazed at the number of new tents that had gone up since I'd left.

A new tent stood by the house spring, and there was Rose, packing a bucket of water up the slope. She smiled at me in recognition and then disappeared into the tent. Yellowstone shied at the tents, and I almost fell off, but by the time Rose came back out I had righted myself and rode past her, straight in that old Porter saddle as though I were a real hand.

I wanted to stop to talk to her, but the cowboys were sitting on the front porch of the bunkhouse, and I knew they'd never stop teasing me. So, despite my long trip from the Hatcher Ranch, I rode up the hill into the pine trees, making my horse jump logs as I entered the woods. Old Yellowstone must have thought I was crazy. Rose was camped on the place for haying season! At least I'd get to see her at a distance. I returned well before dark, but the tent appeared to be empty, and the flaps were tied tight from the outside.

Left to my own devices, I would have camped by the tent until she came back, but I could see my uncle's Chrysler parked at the house. While I was unsaddling Yellowstone in the barn, I heard the sound of hooves in the corral, and Buster Griffin trotted in, leading a bunch of fresh draft horses from the BK to be used in haying.

By the time I had finished helping Buster with the draft horses, everyone was asleep in the big house. I crawled into bed and lay there eager for a new day to begin. I heard a few owl hoots and the call of a nighthawk, and then dropped into a deep sleep far beyond dreams.

All I did that first week was drive a big gray team hooked to a mowing machine, but I loved every minute of it. The horses, Rock and Steel, did all the work, and I sat immersed in the whir of the Pittman and clatter of the teeth on the mower bar as they knifed through the tall timothy and clover. One day I was dreaming about Rose when Ern Morgan shouted at me and shook his fist. I looked behind me to see that I had failed to overlap with the mower, and the meadow looked like the head of a kid with a bad home haircut.

Mowing was hard work for the horses, and every few minutes I stopped to let the sweating animals rest, backing up the team a foot so they did not have to start up with a loaded sickle bar. I couldn't rest long, for ahead of me were a dozen Indians driving mowers. We tried to keep pace with each other, stopping to rest in unison. Round and round we went until the whole field was mowed and the grass lay flat to be cured in the sun, then raked into windrows.

At noon we stopped for lunch, unhooked the teams from the mowers, and led them off to drink from a round spring welling up at the edge of the timber. While the teams lazed in reveries known only to them, we sat on fallen logs and ate, or covered our faces with our hats and napped.

If mowing were all there was to haying I would have loved it, but one morning I found an Indian boy about my age driving my team. I was sent on a crew to cut a derrick pole. Ten of us rode a special horse-drawn wagon with a heavy log X-frame on each end. We traveled up into the pine forest and cut a heavy fifty-foot log, which we peeled, then muscled up until it rode on the X-frame of the wagon. Next we cut and peeled a shorter pole to be used as a spar and loaded it on the floor of the wagon below the derrick pole.

We had almost made it back to the hayfield when the wagon lurched through a small ditch and flung the derrick pole through the air as the wagon tipped over on its right side. There was a sickening sound, like a pumpkin being hit hard by a baseball bat, as the huge log smacked one of the men in the head. The rest of us were thrown over the top of the log and lit on our feet as we scampered out of the way.

The injured man was a white man from Kansas named Harold Nehouse, and the Indians didn't seem to care if he lived or died. They dragged him out of the way, unhooked the team, righted the wagon, and then hooked the team back up. Within minutes the derrick pole was back riding the X-frame, and we were headed out for a piece of dry ground Ern Morgan had picked for a haystack.

It took the remainder of the afternoon to attach a swinging boom to the derrick pole, rig it with cables, and raise it to tower over the stack site. The hay we had mowed the day before was now sun-dry and crisp. The next morning, the men raked the flattened hay with sulky or dump rakes with long, curved tines, which gathered the dried grass and dumped each load in line to extend a windrow. These long ribbons of hay were then bucked into piles with buck rakes or sweeps, long wooden poles six to eight feet long capped with steel, sliding along the ground in front of teams of two horses. The rake teams were always the best teams, capable of pushing a heavy mound, shoving the hay over pole-and-chain nets set on the ground beside the stack, then backing out from under the loads.

Once the nets had been fastened around the hay and closed with a trip latch, the net man signaled the driver of a pull-up team, who drove away from the stack and pulled the load to the top of the derrick pole. The stack foreman swung the load over the stack with the boom spar until it was positioned where he needed the hay. "Dump her!" he called, the net man jerked the trip rope, and the net flew apart, dropping the load of hay on the stack.

Stacking hay on a hot, dusty summer day was no place for a thirteen-year-old kid, but that was where they put me. I was as far away from the horses as I could get. Overwhelmed by heat, choking on dust, I wallowed around in the soft hay, trying to get my legs under me. Ern Morgan could thrust a pitchfork into a whole dump load of hay and shove it right where he wanted it, building the edges of the stack as straight and even as the walls of a brick house. I sweated buckets; hayseed, perspiration, and dust made rivulets down my bare skin and made me itch. At lunchtime I was too exhausted to come down to eat and was still there fast asleep when the foreman dumped a whole load of hay on me.

Little by little, the stack rose in the air. From my vantage point high atop that giant loaf of bread, I could see the buck rake teams pushing loads in from afar, the mower teams mowing tomorrow's hay, and the sulky rakes raking the dried hay into windrows. Harold, the man who had been hit by the boom, was working at a soft job, pulling the nets back to the ground with a saddle horse. He had a bandage around his head, and there were pink roses on the cloth where blood had seeped through.

Midway in the afternoon, as a load was being hauled skyward, a cable broke from the strain, and the frayed end whipped past me and caught Morgan's ear, cutting it badly. He tied a bandanna around his head and went to splicing the cable as though nothing had happened. I was finding out that there wasn't much Morgan couldn't do. A hundred tons were now in the stack, enough to take fifty cows through the winter.

By now my hands were blistered and stuck to the pitchfork handle. My eyes were swollen nearly shut and crusted with dust and grass seed. If I didn't move fast enough in the loose hay, Morgan seemed to delight in dumping the loads on my head. I watched the sun, praying it would go down.

In the distance I saw Rose riding my horse Yellowstone, taking canvas water bags out to the men driving the teams. I struggled even harder, hoping that she would see me there high on the stack, moving hay like a grown man. Morgan seemed to notice that I was trying harder and followed my gaze as I looked over the fields at Rose. A crooked smile turned at the corner of his lips, but he left me alone to my misery.

That afternoon, Buck arrived from town with a new cook and gave Mrs. Biddle strict orders to stay the hell out of the kitchen. There was a hay crew to feed, and if that cook got mad, he said in no uncertain terms, it would be Mrs. Biddle and not the cook who went down the road.

That night, the big oak table in the Yamsi dining room was loaded with enough food for an army. Perhaps from his winter among the Crees, he loved Indians and could not have put up hay without them.

While some of the Indians showed a smoldering resentment toward whites, they treated Buck with respect and only laughed instead of killing him when his bold tongue occasionally pointed out a truth they didn't want to hear. I remember that first haying supper: the table was crowded with Indians, all tired from a long day's work in the hayfields. Buck sat at the head of the table and looked down at the silent faces. "You guys," he said. "You guys are a bunch of murderers! How many killings have you got between you?"

I thought my short life was about to be terminated. I

jerked my feet over the bench, spilling hot soup down my shirt, and got ready to run for it.

An astonished silence hung in the air, and then the Indian on Buck's right started to laugh. "By gawd, Buck, you're sure as hell right. I killed four I know about. Remember that trapper and his kid they found chained up and froze to death in that cabin on Klamath Marsh? That was two of 'em. We'd been partyin', and things got pretty drunk. I'd had a few traps set out with theirs, and I knew they had cheated me out of some mink and marten. When they happened to pass out, I chained 'em to the bed and left. Hell, it was twenty below that night, and when they found them come snowmelt, they was still stiff as boards."

The man next to me wiped some of my soup off his sleeve. "I got a few myself," he admitted. "They all had it comin' to 'em, an' I never did serve time. I remember I had a girl frien' named Edna. She was Shoshone from Fort Hall, Idaho. I just got her so she was pretty handy in bed an' she left me for a white logger over to Bly. All they ever found of that bastard was his fancy cowboots an' his fishin' pole along the Sprague."

I sat big-eyed and quiet as a mouse as the men at the table proceeded to acknowledge twenty-three killings.

As I left the table, a big Indian named Elmer with a scar across his Adam's apple traced the wound with one forefinger as he grinned at me. "You didn't hear a word, kid," he warned.

But I did hear, and those stories stuck with me many a year.

The next day, my uncle hired a man just out of the

Oregon State Penitentiary who was related to some of the Indians on the crew and feared even by his relatives. He was a big man with a knife scar that matched his grin from ear to ear. "Best man with a team of horses I ever saw," Buck said as though justifying the man's presence on the crew.

Margaret Biddle had taken to ignoring my hard work in the hayfields and would put me to work cleaning and dusting whenever she had me alone in the evenings. Needing sleep, I took to staying in the bunkhouse with the men. My purpose may have been twofold, since I was getting a good education in the facts of life, listening to stories from men who had been everywhere and done everything, including murder.

That evening, I managed to catch Rose at the spring and took the heavy bucket of water from her hands and carried it for her toward her tent. I had hardly taken up the bucket when the new man stepped from behind a willow and grabbed the bucket from me. "You stay away from that girl, see!" he snarled.

Rose's face flamed with embarrassment. She grabbed the bucket from the man's hand and fled into the tent.

That night, I lay awake on my iron cot in the bunkhouse trying to sleep. Someone in the darkness across from me began to snore loud enough to make the chimneys on the kerosene lanterns rattle. I stood it as long as I could, then groped for one of my boots and hurled it in the darkness to wake the offender. My boot caught the ex-con on the point of his nose and split his flesh right to his hairline.

Someone lit a lantern as the enraged man, covered with blood, charged about the room in his underwear, ready to do violence with whoever had flung that boot. I was the one with only one boot under my cot. He seemed ready to blame the old chore man in the cot next to mine, however, before I spoke up. "I did it!" I said. "I threw that boot. You were keeping the whole damn bunkhouse awake with your snores."

There was a long silence, as though the man was remembering that this was the insolent pup who had taken a shine to Rose. He took a lantern, lit it with a big kitchen match, and walked to the mirror to inspect the damage. Suddenly he began to laugh, and the tension in the bunkhouse melted like butter. "Dang!" he said. "Dang if the kid didn't improve my looks."

The next day it rained too hard to hay, and the foreman paired me with the ex-convict to take out a wagonload of livestock salt to the cattle. For a time the man's face was frozen as though it hurt him to smile. I kept both hands on the Jacob's staff, a notched upright on the front of the bed where the teamster ties his reins when leaving the wagon. I was ready to hurl myself off and run should he make a move.

It wasn't long, however, before I forgot my fears. I kept watching the man's hands on the lines and how well the horses read his wishes, as though some mysterious thought process traveled down those leather ribbons. My uncle was right. The man had a great way with a team.

"Sure wish I could handle horses like that," I said.

For a long moment the man stared at me, and then he grinned. "I reckon there ain't no better time than this to learn," he said. Handing over the lines, he stepped into the air and headed back to camp.

Chapter Five

THE ONLY THING THAT COULD INTERRUPT a haying operation was broken machinery or rain. If rain fell on our hayfields and not the neighbor's, we went to help them, but if the rain was general, we got the day off to heal.

One morning, I was awakened early by the roar of a heavy mountain rainstorm pounding the shingles of the ranch house roof and knew that it was too wet for haying. I could have slept the clock around, but hunger gnawed at my stomach, and I dressed and slipped down to the kitchen. The old lady was off traveling, and my uncle was sitting at the kitchen table reading *Time* magazine. He wore his favorite wool cardigan with leather patches on both elbows and seemed to sense my presence without even glancing my way. "Yeah," he grunted by way of greeting, and went on reading.

He got up from his seat, went to the stove, and poured a cup of tea. I guessed it was for himself, but he set it before me, gave me a pat on my shoulder, and resumed his reading. That one little gesture brought tears to my eyes. Underneath that solemn exterior, he cared! Not being able to hear, maybe he compensated. Instead of conversations, he spoke in monologues, and when you were with him you listened rather than try to interrupt. Maybe, I thought, maybe he really likes kids but with his hearing problems doesn't know how to talk to them.

I fried myself a couple of eggs, made an egg sandwich with a cold biscuit, and sat down at the table. "Uncle," I said, putting my hand on his arm for attention. "My mom said you were the best trout fisherman to come out of Michigan. If I really tried hard to learn, would you teach me?"

My uncle did not smile or comment. Instead he rose from his chair, left the kitchen, and returned with a couple of rods, one of which he handed me, and we were off in the rain to fish the river.

The old man had a strange way of fishing. He used a long cane pole and a short line to which was attached a tiny silver spinner. He never cast more than three times in any of the deep holes at the bends of the meandering stream but moved off, restless to be on. Soon he was out of sight in the mists, leaving me to my own solitary ways. It was almost dark when I saw him again. He had traveled north fifteen miles and had only a couple of trout to show for it. I had stayed at the first bend of the river and caught twenty.

That night we feasted on Yamsi trout, dipped in cornmeal and fried in butter. From that day on I never again saw him fish, but he took me often, and sat quietly on the bank and watched as I reeled them in.

For the next week, storm after storm swept in over the Cascades and kept us from putting up hay. One rainy day, as I rode BK Heavy past Rose's tent, she happened out on her way to the outhouse.

She looked at me and smiled shyly. "Hey," she said. "You

want to go over to the Sycan River with us to catch floppers?" Her long black hair wasn't braided yet this morning. Half fell over her right breast, and the rest cascaded down over her shoulders. Raindrops glistened like diamonds on the strands.

"Sure," I said. "Let me put this old horse away and get on a dry shirt." I had no idea what floppers were, but I was burning up with excitement at spending some time with her.

A half hour later I was crowded into the back of an old pickup truck with a dozen half-drowned Indian kids, bouncing up over the ridges of Taylor Butte toward the Sycan River drainage. It turned out that floppers were young ducks that had grown to full size but had not yet acquired the power of flight.

The sloughs and potholes along the Sycan River were full of them, and we scampered shivering through the rain-drenched sloughs in water up to our hips, catching the largest of the ducks and loading them into the pickup until we had enough for several good meals. I was a little bit disappointed. Even though Rose had asked me to go along, she ignored me to hunt with the girls, while I was surrounded by boys.

This was reservation land, and Indians could hunt and fish as they pleased without worrying about legal methods or seasons. I was conscious that I was a white kid, and kept looking back over my shoulder expecting that any moment now a federal warden would cart me off to jail.

On the way back to the ranch the woman who was driving slammed on the brakes and stopped the pickup. Grabbing a rifle from the window rack, she shot a big buck that had crossed the road in front of her. The antlers of the deer were covered with velvet, and it was fat and sleek. The woman made short work of gutting out the animal and piled the entrails alongside the road as a signal to other Indians that hunting had been good here.

"We'll make some good jerky," Rose said, addressing me for the first time in an hour. "You like jerky?"

I shrugged with the indifference of a teenager. I was paying her back for ignoring me.

We ended up at a deer camp just outside my uncle's fence at the head of the Williamson River. There were pickup trucks everywhere, all equipped with what was standard equipment for the Indian, a spotlight atop the cab for night hunting and a rack of rifles in the back window. Most of the men had been hunting all night and were sleeping, but there were several sitting around the campfire drinking beer.

They watched with mild interest as the old woman who had shot the deer slid it from the pickup and skinned out the animal on the ground. No one made any effort to help. The woman worked swiftly and soon had rendered the whole carcass into thin strips, which she draped over drying racks of poultry netting stretched over the fire.

She nodded to Rose, and said a few words in the Klamath language. Rose dragged the fresh hide over to my uncle's fence and draped it raw side up over the barbed wire to dry. There were at least a hundred other hides hanging there; some were so tiny they obviously had come from small fawns.

Smoke rose from the willow fires and bathed the drying

venison, but there were bluebottle flies buzzing everywhere, flying through the smoke to lay their eggs on the meat.

Rose helped the woman scatter pepper on the jerky strips, then took a piece off the rack, shook off a couple of bluebottle flies, and offered it to me. I shook my head. I was dying to try some, but the flies made my stomach squirm. Rose had no such qualms. Her cheek bulged as though she were storing a chaw of tobacco, but it didn't interfere with her speech. "You goin' to the rodeo tomorrow in Beatty?" she asked.

"Where's Beatty?"

"On the Res. 'Bout twenty miles south of here."

"That far? I guess not. I got no way to get there. And if it's dry enough I'll have to work in the hay."

"Your uncle won't have a crew," Rose said. "All us Indians are going to the rodeo. They say Jack Sherman's going to make an exhibition ride on Blackhawk. I tell you that old black horse can buck up a storm."

She moved the wad of jerky to the other cheek. "You could ride over with us," she said.

I was so mesmerized by the pretty roundness of her face and the dark mystery of her eyes that I scarcely heard her.

"You could ride over with us," she repeated.

That night I could hardly sleep for thinking about going to the rodeo with Rose. But in the morning, Buck woke me early and insisted that I ride along with him to open gates. As we left to drive down through the ranch, I saw Rose and her family loading up for the rodeo. I sat hunkered in the front seat, overwhelmed with self-pity, hoping that the raindrops misting my uncle's windshield would turn into a deluge, hoping he would get stuck in the mud, that he would run out of gas, that the engine would fail to start.

We had gone less than a mile when the old man thought of something he had forgotten to do in town and made a U-turn, dumping me back at the house. The old lady was there and calling my name. Not wanting to get caught up in housework, I stayed outside in the shelter of an overhang until I heard the distant sound of his Chrysler as he sped up the first hill. I was too late to catch a ride with Rose, so I fled to the corrals, captured old Sleepy, and was soon trotting south over the hills toward Beatty, twenty-three miles away.

Sleepy had one redeeming feature in that he would rather trot than walk. He moved out eagerly, singing his groaning complaint but eating up the miles. I rode standing in the stirrups, looking out for the black-and-white signs with which the Indian Agency marked the roads. I soon left the timber behind, and passed over miles of rocky flats covered with mountain mahogany and small groves of aspen. Five miles from town, as I crossed a vastness of sagebrush, I began to see clouds of dust from the south and knew that the rodeo had already started.

The arena lay just south of town. The surrounding fence was pretty primitive, but it was buffered by cars parked fender to fender and crowds of people. Here and there Indian women sat on blankets, legs extended, playing at gambling games, hiding pieces of painted bones in their hands. The cowboys in the arena were mostly Indian with a sprinkling of whites. Behind the parked cars were hordes of children riding horses, galloping back and forth, raising more dust than the cowboys. They all seemed to be riding with one hand on the reins, the other with a death grip on a bottle of pop, and were much more interested in each other than in what was happening in the arena.

Rose spotted me from afar and came running, vaulting up behind me on Sleepy's back as the old horse threatened to dump us both. He crow-hopped a few jumps then settled down to a walk. Now and then I felt Rose's breasts bump my shoulder blades and I was in heaven.

We found a place along the fence not far from the bucking chutes and sat astride Sleepy watching the rodeo from our vantage point. I had missed the bareback riding event, but the saddle broncs would come later. In the arena was a confusion of ropers of all sizes, shapes, ages, and degrees of inebriation. Some of them fell off their horses before they caught up with the animals they were trying to rope.

"That's my uncle," Rose said as a plump Indian fell off his horse and lay still. Another rider roped the man by the feet and dragged him out of the arena to laughter all around.

"Why do they drink so much?" I asked Rose.

She shrugged. "I guess they just like being drunk," she replied. There was a loud clatter from the bucking chutes as the Indians began to fill the chutes with big, stout horses.

"Those are the saddle broncs," Rose whispered in awe. "The big black horse in chute number one is Blackhawk.

Dally Givons says he's the best there is. Jack Sherman's going to try to ride him. Bart Shelley owns him, an' every rodeo contractor in the country would like to buy him but old Bart won't sell."

In an effort to see around me, Rose put her cheek close to mine and a lock of raven hair fell down over my chest. "See," she said. "There's Jack now over by the chutes, sitting on his saddle on the ground, getting his stirrups set for his ride."

Jack was a tall, angular cowboy, with sandy hair, gray eyes, and a crooked smile. He was a natural athlete and, though I didn't know it at the time, was one of the great bronc riders of the forties. I began to worship Jack as a hero before I ever saw him ride.

Jack climbed the chute beside the big horse and lowered his bronc saddle on its back, then used a long wire with a hook to draw up the cinch beneath the animal's belly. Blackhawk snorted as the cinch came up tight and hammered the planks of the chute behind him with jagged hooves. I watched mesmerized as the big man strapped a braided rein to Blackhawk's halter, drew it back to the swells of the saddle, and marked a place on the rein with a wisp plucked from Blackhawk's mane.

"What's he doing that for?" I asked Rose.

"He's marking a place to hold the rein. Too short and the horse will duck his head and jerk the cowboy forward out of the saddle. Too long and the rider can't keep his seat."

"You think he can ride him?" I asked, a charge of excitement running up my spine.

"Who knows?" she replied. "On a good day, Dally Givons says, Jack Sherman can ride anything with hair."

The crowd went utterly silent as Jack climbed the slats of the chute and settled down on Blackhawk's back. The big black turned his head around as though to see who dared to try him. His black eyes snapped in anger as he waited for the side of the chute to open.

"Watch closely," Rose hissed. "To make a qualified ride, Jack has to keep his spurs in the horse's neck through the first jump and ride for ten seconds."

Jack carefully gripped his rein where he had marked it. His toes found the stirrups. His eyes glanced out into the arena at the two cowboys, the mounted pickup men, who would help him from the horse at the end of the ride. "Let's have 'im," he said, nodding to the gate man, and the chute gate shot open.

Blackhawk followed the gate out with a wild plunge that knocked Jack's hat off on the planks above the chute, then bucked high and wild, turning his body flat in the air in what is called a sunfish. For a moment it appeared the big horse would fall flat on his side, crushing the cowboy's leg, but Blackhawk caught himself and threw in a couple of crooked, jolting jumps. The crowd screamed as Jack Sherman came closer and closer to completing his ten-second ride.

A big woman lurched against old Sleepy, almost knocking the animal off its feet. "Jump high and fart loud!" she shouted, spilling her cup of beer all over my pant leg. But suddenly the big horse leaped high, hit hard, and sucked back beneath himself, sending Jack catapulting off. Jack lit on his shoulders in the dirt and rolled away from the horse's hooves. He lay for a moment watching Blackhawk buck on without him, empty stirrups hitting above the saddle, then got up slowly, a crooked grin on his lips, took a sack of Bull Durham tobacco from his shirt pocket, rolled a cigarette, lit it, and limped back to the chutes. Maybe it was just a little Indian rodeo at Beatty, but I sensed that rodeo didn't get much better than that.

That night, a full moon lit my way across the tablelands. Sleepy trotted on and on, following the slender ribbon of a road as it wandered through the pines. On my left I could make out the familiar landmark of Fuego Mountain; on my right I could see a gap in the trees formed by the steep rocky canyon of the Sycan River. As I rode up the ridge toward Taylor Butte, I could feel the icy pockets of cold air settling in the hollows. Ahead of me, a great gray owl caught a mouse and sat on a stump until my horse got too close, then flew off into the darkness with its prize.

Now and then I would see sudden flashes of lights towering skyward as a carload of Indians spotlighted deer. Sometimes the beam would illuminate me and old Sleepy, and the hunters would slam on their brakes and the lights would go out as they slipped on past and the rumble of their vehicle was lost in the pines.

It was after midnight when I finally made it back to the Yamsi barn. I groped about in the darkness to unsaddle Sleepy and then turned him out to roll in the corral. In the distance one of Sleepy's friends nickered to him, and the tired bay horse answered back. As I walked toward the bunkhouse, I heard the hollow drumming of his hooves crossing the plank bridge, and then splashing sounds as the horse trotted off across the wet meadow to join his friend.

Chapter Six

By the time I was sixteen, I had grown into a six-foot-five beanpole of a cowboy. Some folks claimed I kept my pockets full of rocks to keep from blowing away. Strangers would grin at me as I walked the streets of Klamath Falls and say, "Hey, how's the weather up there, Slim?" I longed to spit in their eyes and retort, "It's rainin' up here. How is it down there?"

But I didn't, of course. I just slouched on my way, hoping to duck around the next corner and disappear. I even daydreamed of becoming invisible. But invisibility was a tough trick to pull off on Main Street. In those days, I couldn't walk a block on Main Street without passing several Indians off the reservation, people who needed me as much as I needed them. They were my solace, for I could stop and talk to them about horses and rodeo, and they treated me like a friend rather than a freak. By stopping in at Charley Read's saddle shop, I could count on seeing some Indian friends like Buck Scott or Lee Hutchison, only a little older than I, who were already becoming pretty good saddle bronc riders.

Or I could go over to the Montgomery Ward saddle department, where Jerry Ambler, the reigning world champion saddle bronc rider, ran the department. Jerry was slender as a willow whip and rode by balance instead of brute strength. He sat the saddle on a space no bigger than a handkerchief, shoulders bowed, looking as though he were being towed along by his bucking rein. Of the balance riders I watched through the years, probably only the great South Dakotan Casey Tibbs, from Fort Pierre, was his equal. Maybe Jerry could sense my intense interest and knowledge of bucking horses, for he was always nice to me and would drive out of his way to pick me up at the ranch and take me to a rodeo.

In those days, before World War II, a great many rodeo contestants were ranch-raised kids who had grown up a-horseback and learned to ride bucking horses by breaking colts. Perhaps they rode the rough string and spoiled horses for a big outfit like the ZX over at Paisley. They might go to town for a Fourth of July rodeo and compete against other riders from other ranches, take home a little prize money, and go back to buckarooing on the ranches. If they were good enough, they made rodeos on weekends, paid their entrance fees, and hoped that the prize money they managed to win would exceed gas money, food, entrance fees, and doctor bills.

I read and reread the rodeo magazine *Hoofs and Horns* until the pages fell apart. My heroes were the bronc riders of the day — Bill McMackin, Doff Aber, Perry Ivory, Gene Rambo, to name a few. In my dreams I competed at Madison Square Garden, Pendleton, Calgary, and Salinas. I rode high, wide, and handsome until they opened the chute gate and my dream mount plunged bucking and kicking into the arena. There the dream always ended, for I could not fight reality and was suddenly snapped back to being a tall, gangly kid, a wannabe with no talent for riding the really tough horses.

My head was full of great bucking horses like Steamboat, Tipperary, Midnight, and Five Minutes Till Midnight, and the bunkhouse cowboys must have gotten sick of hearing about them.

Each of the cowboys at Yamsi had six horses on his string, one for each of the six working days. I was no exception and gloried in those that might buck a little. I had a big gray horse named Smoky that was cold-backed, meaning he would hump up and buck, in the morning when I got on. We were a pretty good team. Smoky liked to buck and I liked to ride. Without any encouragement on my part, Smoky would drop his head between his knees and buck in long, easy jumps. He leaped high, and made me feel more talented than I really was.

I wanted desperately to belong somewhere, to earn the respect of those I idolized, but I was still growing like a weed, and no matter how much I wanted to sign up as a contestant at each rodeo I went to, I ended up just hanging around the chutes, watching for any opportunity to help. There were lots of rodeos each summer, and I went to every one I could. Beyond the rodeo events, there were parades and dances. The towns filled up with local ranchers and cowboys, and for many of them it was their one trip to town for the year, the source of many a tale to be told again and again around the bunkhouse stoves.

One Sunday, I rode over to the Beatty Rodeo with Rose

and her family. On the way, we passed Bart Shelley, driving a bunch of his horses across the vast sagebrush flats toward town. I was thrilled to see the big horse Blackhawk in the lead. I had brought a camera with me and couldn't wait to photograph Blackhawk coming out of the chute. He was ridden by a cowboy named Ed Donovan, who managed a ten-second ride. I took the film to a Klamath Falls drugstore for development and sweated out the results for nearly a week. With most of the bunkhouse cowboys gathered around me, I opened the envelope from the photo shop and found I had captured Blackhawk sunfishing high in the air. It was to be the first of thousands of rodeo photographs I would take through the years, and that gave me a niche in rodeo beyond being a mere spectator.

I had no formal training as a photographer but learned by doing. Most of my early attempts ranged from bad to awful, and it took several rodeos before I shot anything as good as the photograph of Ed Donovan on Blackhawk.

Soon I was sending off a flood of rodeo shots to Ma Hopkins, the editor of *Hoofs and Horns,* and it wasn't too hard to persuade her to give me a job as official photographer. She neglected to ask me my age and would have been aghast had she known I was only sixteen.

Without rodeos, Saturday nights at the ranch were lonely, and often I rode into town with some of the Yamsi Ranch crew and sat in the darkness outside houses known as Irene's or the Iron Door, as the cowboys sought out what they had come to town for. I would listen and wonder as music and laughter came from within. Sometimes a woman would come out to bring me a bottle of pop and sit in the cab of the truck, visiting me. I would marvel that these women seemed like any other women I ran into in town. They talked of families in towns I'd never heard of, and what they dreamed of doing with their lives once they got money. They called me "honey" and "sweetheart," and their perfume lingered on long after they had gone back to the house.

On the ranch, on long days a-horseback, I did lots of listening, for I felt at my age I had few stories worth telling. What I was learning, in those days, was to be a good audience. I honored the storytellers by being attentive, by riding carefully abreast of them so as not to miss a single word.

Life on the ranch, of course, was not all stories of the past. Drama was a part of everyday life. One summer day Ern Morgan sent me off to help on the Marsh with an old cowboy named Roy. We shunned the roads and cut across country, both of us tired of the usual well-worn trails.

Roy was up in his seventies but fighting old age as though life had gone too fast and there were still places to go and things to do. I remember listening to the creak of him as we trotted along through towering virgin ponderosa pines, leaving a faint trail of pumice dust as we rode. I wasn't sure whether the creak came from his old bones or from his ancient Hamley saddle, which showed a lifetime of hard use. We were conscious, both of us, that one day these giant trees would be cut and hauled away, and we had best drink in the scenery before it was gone.

I had the usual pity of youth for the old, after seeing that Roy needed a pine stump to mount and having heard him groan as he settled into the saddle. But before the day was out I learned that the old man had plenty of life left in him.

That afternoon we were dropping down off the ridges to the edge of Klamath Marsh, along Big Wocus Bay, when the old man's horse stepped on a stick and began to limp. We still had miles to go, and there seemed to be little we could do but spend the night in the woods. But Roy had other ideas.

"Look," he said, pointing to fresh tracks in the pumice and a faint haze of dust still hanging in the forest. "There's a band of wild horses just ahead of us heading down to water. Maybe I'll just get me another horse."

We moved to a rim above Klamath Marsh, and sat and watched the horses from afar as they came out of the forest, crossed a meadow, and moved out toward the shine of water. The old cowboy crawled off his horse and tightened his cinch, motioning for me to tighten mine. "Can you rope, kid?" he asked.

"Just a little," I replied. "Damn little, if you want to know the truth."

Calmly, he took down his rawhide reata and shook out a loop so big it almost dragged the ground.

"Get ready to ride, boy. I'm about to catch me a fresh horse."

He ignored my look of disbelief. "See that little peninsula that sticks out into the marsh? I reckon they'll head out there to drink, then we'll rush 'em. If it works, I might be able to get in a good throw as they try to break past us. Reckon they'll do that rather than chance getting stuck out there in that muck."

I stared at the old man, thinking he was crazy, but suddenly we were out of the woods and the spooked horses were blasting on ahead, running right out on that narrow neck. The old lead mare reached the end, paused, looked back, then piled into the water, throwing up a spray that drenched the other animals. She panicked in the muck and lurched back out of the water onto firm ground.

"Get a loop built, kid!" Roy shouted. "Here they come!"

Before we were ready, the wild horses saw the trap and charged for freedom. Suddenly they were all around us as they tried to escape. I threw a desperation loop at a young sorrel and caught a willow bush instead. I looked around just in time to see Roy drop a loop around the big, battle-scarred gray stud, jerk his slack, and take his dallies around his saddle horn. "Git around him, boy!" Roy shouted. The rubber covering on his saddle horn began to smoke and stink as he rendered slack in his rope. The stallion slowed and plunged, nearly jerking Roy's horse down. Giving in to the pressure of the rope, the desperate animal reared and fell over on his back.

As I galloped between the stud and the vanishing herd, I built another loop. The horse came to its feet, reared, and strained, trying to break the reata. Through flaring nostrils it gasped for air. I laid a loop in front of his hind feet as Roy dragged the horse into my trap. Stretched between two ropes, the stallion screamed in anger, then tumbled to the ground flat on its side. Keeping his reata tight, Roy rode around a small pine tree, stepping his horse across the rope three times. Then he slipped from his horse and tied the reata to the pine. "Ain't the hoss I throwed at," Roy admitted, "but he'll do.

"Keep your rope tight!" he ordered as he unsaddled his horse.

Approaching the angry animal carefully, he took his tie rope by both ends and sawed it under the horse's girth, then, as the horse snapped at him with yellowed teeth, got his old Hamley. Tying one end of the tie rope to the cinch ring, he pulled the cinch and latigo underneath the horse until the saddle sat in place on the animal's back. Soon the saddle was cinched tight and ready for a rider.

The old man was out of breath. For a few minutes he sat on a log to catch his wind. I had the desperate fear that he was going to make me ride in his place.

"You think I'm going to ride that horse, you're crazy as hell," I snapped.

"You ride him?" Roy grinned. "Why, hell no. I'm going to show you a thing or two. Being young or old doesn't matter half as much as knowin' how."

Roy straightened himself up to all of his five feet four. I could hear his old frame snap as he moved. "When I'm good and ready, you're goin' to turn 'im loose. You just be damn sure you keep 'im herded out in the mud away from shore!"

I was full of can'ts at that point, scared to death of riding in that quaking morass of a stinking black swamp. But before I knew it, Roy had jerked the snaffle bit off his own horse and had it between the stallion's yellow teeth. He knelt on the horse's neck while he got the reins in place and the throat latch buckled.

"Git ready," he ordered. "Give 'im slack when I tell you

and let 'im have his hind feet. We'll pick up my reata tomorrow when we come back this way."

He loosened his reata and tossed it free of the animal's head, then crouched over the saddle. With a lunge, the horse came to its feet. Roy found the off stirrup with his toe. "Wish me luck, kid!" he shouted.

With a scream of anger, the stud leaped toward the marsh. Out in the deep the horse lost its footing, and for a moment there was only the boil of brown swamp water. Then, suddenly, both horse and rider were up and sputtering. Roy pulled the animal's head around until it was pointed parallel to the shore.

"Haaaah!" I shouted, kicking my horse into the water.

One more jump and the pair went down again, but this time the angry animal kept its head up. Finding its footing, it lurched forward through the mud. Every time the horse would lower its head to buck, its nostrils would go under, and it would lift its head and charge on. Once it reared and fell sideways, pinning Roy under, but the man pulled the horse's nose beneath the surface and it struggled up again to its feet. I kept busy splashing back and forth, keeping the animal herded away from shore.

We had gone only a quarter of a mile or so when Roy sensed that the animal was ready. "Ride right in front of him," the old man said. "Lead him ashore. I think he'll follow your horse."

Before I knew it, Roy and the horse were on firm ground, and the stud, which had wanted to kill a cowboy only moments before, now trotted along behind mine, with only an occasional look back at the forest where his mares had disappeared. We moved west along the shoreline, holding as close as possible to the swamp, with Roy's old saddle horse bringing up the rear. I shook my head in disbelief. I had seen a real cowboy at work and seen a touch of the past when old men were still men.

That fall, Roy and I were staying in the old white Houston ranch house on the west side of Klamath Marsh, riding for strays. The land was white with the first snows off the Cascades, towering above us to the west. I had come in early to start the cook fire, and Roy was taking one last sashay out on the flats to check stock water in the troughs. Outside the house I heard hoofbeats and looked out to see Roy's horse coming in alone. The old Hamley saddle was empty, and the reins were dragging. Roy had been one of the lucky ones, avoiding old age. I got the neighbors, and we brought his body in by moonlight with a team and wagon.

Chapter Seven

UNTIL WORLD WAR II CAME ALONG, changes to the West came slowly, almost imperceptibly. Then, almost overnight, the young men were gone, and even older ranch hands with able bodies had gone off to make money in factories. The ranches were left to survive as best they could with the dregs of the labor pool.

At Yamsi, we used old saddle horses that should have been retired, since there was no one around to ride Whingding or break colts. The young horses were there, of course. You can't turn off a pregnant mare. But the horses that were ready for training got older every year and harder to handle when someone did pass by willing to start one.

At seventeen, I wasn't quite old enough for the service, but I had to grow up in a hurry with all the responsibilities that were thrust upon me. I was a-horseback dawn to dark, moving cattle, doctoring, looking for strays. Buck had to open his own gates, for I was gone before he got up in the morning and back after he had gone to bed.

For years I had wanted to ride Whingding, but he had always been in someone else's string. Now he was mine if I wanted him. He snorted a little as I got on him one cold morning, but he seemed to know I was itching for him to explode, and I couldn't have made him buck if I tried.

My uncle liked to migrate to warmer climes in winter. Rather than be stuck with me, he shipped me off to school in California. I was an indifferent student, my mind filled with horses rather than mathematics and Latin. I couldn't wait to get back to the ranch for Christmas vacation.

Most of the crew had been sent to the Williams, California, area to take care of several thousand wintering cattle. I spent the first of my vacation at Yamsi on long, bone-chilling rides to gather strays from lonely draws or meadows, which, coated with snow, were a far cry from the beautiful grasslands of summer. Outside of Ern Morgan, an old sheepman named Jim O'Connor, and Fred Shepherd, who we called Shep, the ancient chore man, there was no one around to help when things needed being done.

Morgan was beside himself with worry. "This goddam war! All the employment office in Klamath Falls ever sends me are winos, stinking of strawberry wine. I'm already three weeks late getting the cattle out of here for the winter, and if a storm strikes we're in deep trouble!"

There was just so much hay in the Yamsi haystacks, but the long meadows were dark with cattle gathered off the ninety-thousand-acre range, cattle that would soon have to be fed. Morgan was right. We should already be moving out with the herd, driving them overland across the tablelands to the wintering area on the BK Ranch at Bly.

Morgan stayed at the telephone, trying to find men, but when the promised help failed to show up, the old foreman had little choice but to pray for decent weather. Each day we delayed increased the chances we would be snowed in with the cattle for the winter. Every day found Morgan up at dawn's first light watching for clouds scudding along the tops of the Cascade Mountains to the west, indicating whether or not a storm was in the offing.

It was twenty below zero and still dark when we saddled our horses and started the herd of six hundred dry cows up over the long, snowy trail through the pine forest, up past Taylor Butte toward Bly, forty miles away. Most years the cows would have been eager for the trip and the herd stretched out down that snowy road through the lonely forest, thinking only of the stacks of timothy, clover, and bluegrass hay awaiting them at the end of the trail. But this time there were no leaders. These cattle had been summered on Klamath Marsh and wintered in California. They had no idea of the trail.

We were so desperate for riders, we let the old lady come along to help us over the hills. She was bundled up so only the tip of her nose and her faded blue eyes showed, but we saddled her horse, Ginger, for her, lifted her up on a stump, and shoved her aboard. There wasn't a peep out of her, swathed as she was in wool. It turned out we couldn't have made it up that first hill without her.

We had gone maybe two miles when it started to snow hard, and in minutes the backs of the cattle were covered, making them hard to see. It was just breaking daylight, but the woods were still ominously dark with the storm. Had the decision been mine, I would have turned the herd back, but somewhere in front of us, Ernie Morgan had taken a small bunch of cows at the point and was pushing them up the hill, breaking trail in what was now knee-deep snow. Morgan wasn't a man to give up, and besides that, he knew from long experience that this might be the last chance left to get the cattle out of the valley before Yamsi Ranch was snowed in until spring.

The cows were hungry and stopped at each snow-covered bitterbrush bush to nose the snow from the branches and strip the leaves. I was riding Whingding at the rear of the herd, and he shied and snorted at each unfamiliar snow-covered shape. All I need now, I thought, is to get bucked off and have the drag take off back to the ranch.

All I had for help in the drag was a couple of red-nosed, well-intentioned neighbor kids and Margaret Biddle. Somewhere up along the flank was seventy-five-year-old Jim O'Connor, who would be doing his best because that was Jim's way. But he was known as a better hand with sheep than with cows. Jim wasn't very well mounted, and besides, there was only so much any man could do in a storm like this.

I screamed and cursed at the cattle, but the wind tore my words away. The kids seemed to be trying hard. I grinned at them whenever I passed, but already their horses were leaden with fatigue. Even with daylight, there was not much visibility. Each of us rode in a small world limited by how far we could see around us in that storm. We would ride up to a cow, scream, and pop the animal with the ends of our reins, only to find we were trying to drive a snow-covered bush. Whingding moved angrily back and forth, ears laid back, biting at the backs of cattle to drive them on. There was a gleam in his eye; he was tough as iron, and this was what he had been bred to do. He had two passions in life, bucking off cowboys and driving cattle.

We rode with one hand under our chaps for warmth, but one hand had to be out in the weather to rein the horse. Our fingers ached horribly, and our cheeks burned even as we snuggled our faces in our scarves and rode with heads bowed against the onslaught.

Now and then I would see the old sheepman ahead of me, hunched up and cold but doing his level best. He would glance back at me, reading my misery.

"Be the Jaisus, lad," he'd shout with a grin. "Ain't this fun, though? 'Tis a great day I be havin'!"

Trailing the herd at a distance was the chuck wagon, pulled by the gray Percheron draft team, Rock and Steel. Shep was in his eighties but a good chore man. He had been instructed to follow the herd to Eldon Springs, where we planned to take shelter for the night. The wagon carried a tent and our bedrolls, plus beefsteaks, eggs, coffee, biscuits, and candy bars to help us survive the ordeal of sleeping out in a blizzard.

Margaret Biddle and the kids in the drag were pounding on the cattle, moving them up twenty feet at a time, but as soon as the pressure was off and the riders rode after another laggard cow, the first cows would stop to eat. I trotted back and forth behind the herd, making sure I kept on the outside of all the tracks. Now and then I would find a bunch of cows trying to sneak off through the pines and head back down the hill. I would scream at them like a banshee and run them back into the herd.

It was noon when we finally made it to the top of the ridge, but here the wind came charging down off Fuego Mountain, ready to freeze us to our saddles, and blinding us with icy pellets. I hadn't seen Morgan in two hours and hoped he could still locate the road through the trees. The old lady was a frozen lump on her horse, and I begged her to go back to the ranch. A few teardrops froze to my face, for I wasn't sure I would ever see her again alive.

One of the kids in the drag was too cold to get off his horse and tried to pee from the saddle, but all he managed to do was wet his chaps. The laugh we had at his expense made us feel a little better, but we all had our own battle to fight, and the storm seemed determined to defeat us.

We hit the Sycan River at Teddy Power Meadow. Morgan somehow kept the cattle in the timber, for the open flats were a whiteout of blown snow, and the cattle would have bunched up, tails against the wind, refusing to move. There were trees across the road, felled by the storm, and the cattle had trouble working their way around them. Every delay cost us precious time. The days were short, and we dreaded having darkness catch us still out on the trail.

Three o'clock, and the storm showed no signs of abating. Whingding was tired but still determined to make the cows go. We were out of the pines now and surrounded by groves of mountain mahogany that gave scant shelter.

"It won't be long now," I shouted to the kids in the drag, but they were so bundled up in misery, they didn't seem to hear.

Darkness came, and still we had not come to Eldon Springs. I was worried about the chuck wagon. It had been right on our tail at Teddy Power Meadow, but I had not seen it since. I kept thinking about possible disasters. What if old Fred lost track of the herd in the snow? What if he had a heart attack and died? What if the wagon lost a wheel, or a tree collapsed on the team?

We kept pounding on the cattle until at last the herd became thicker, and some of the tired cattle were even bedding down in the snow. I recognized some of those cows as having been part of Morgan's herd in the lead. Maybe we had gotten as far as he wanted to go. Not far ahead I saw flames shoot up as Ern Morgan set fire to a big pitch snag. We had made it to the shelter of aspen groves at Eldon Springs, and here we would set up camp with the wagon and hot food, spending the night warm and snug in the tent.

Or so I thought. We got down off our horses and crowded up to the fire for warmth. Yellow pitch dripped from the burning snag, and black smoke poured up to bore a hole in the clouds of swirling snow. Steam rose from our frozen clothing. Once we could function, we unsaddled, letting our tired horses roll in the snow. Morgan kept looking back into the darkness the way we had come.

"Somethin's wrong," he said flatly. "Old Shep and the wagon should have been here long ago."

We fed our horses from a stack of loose hay that rancher Bart Shelley had left for emergencies. I volunteered to go looking for the old man and the wagon. Surely they couldn't be very far back on the trail. I started to saddle Whingding, but the old horse stood with his head down, too tired even to eat.

"I'll go afoot," I told Morgan. "He can't be that far away."

At the end of the first mile, I stumbled over the body of a cow that had died unnoticed on the trail. But still there was no sign of the wagon. I rested for a few moments in the lee of a big pine, then decided to go back to looking for Shep. I could hardly make out the road. Here and there the wind had bared the trail. Here and there it had buried our tracks in huge drifts. I started to get frightened that I might lose my way, die on the trail, and never see Yamsi again.

My feet were numb in my overshoes, and I stopped frequently at trees to kick the bark to restore circulation. At Teddy Power Meadow, I lost the road and wandered up on the frozen meadows where once I had chased flopper ducks with my friends. It all seemed so long ago.

The going was easier here where the wind had swept much of the snow away. In the darkness I made out the walls of Teddy's old log cabin and took shelter from the wind to rest.

There were only three sides to the cabin. Teddy's Indian wife had been a large woman, so big, in fact, that when she got down with appendicitis, Teddy had to tear out the end of the cabin to get her out the door. She took up the whole front seat of the buckboard, and Teddy stood behind her handling the reins all the way to Beatty, sixteen miles away.

I seemed to hear Teddy's voice in the darkness encouraging me to go on, giving me advice. "Better go on, kid. You stay here, you won't last the night."

Back on the forest road again, I found signs of the chuck wagon. From the tracks, Shep had passed this way twice. Apparently the old man had lost our trail and headed back to the ranch, leaving us to our fate. Six miles later, I staggered through the door of the bunkhouse to find Shep fast asleep in his bed.

I might have let the old man sleep, but I was mad at him for letting us down. Besides, we were shorthanded and needed his help with the wagon. Grabbing a side of his cot, I dumped him, bedroll and all, on the floor.

I was relieved to find Ginger tied to his stall and see lights on in the big house. Mrs. Biddle had made it home safely. Working quickly, I had a fresh team harnessed when Shep caught up with me at the barn.

"Danged smart-alec pup," he snorted. "I was dreamin' I was chasin' a beautiful redheaded girl and was just about to catch her too when you dumped me out of my bed."

"You were layin' there dreamin' when the rest of us were about to die of hunger out in that storm!"

"Don't know where you disappeared to," the old man said. "That herd of cows just left off leavin' tracks."

Moments later, wrapped in blankets and canvas, we were bundled up on the front seat of the chuck wagon and headed south toward Eldon Springs with food, tent, bedrolls, and hot coffee, plus some rolled oats for all our hardworking horses. There were grins of relief when we finally cleared the last bend and found the little party still clustered around the fire.

We slept a little late in the morning, for the big tent was

swaybacked with fresh snow, and we suspected that today was going to be colder and tougher than the day before. We sat in our bedrolls, dressing slowly, careful not to dislodge the lining of frost that clung to the inside of the canvas from the moisture in our breath.

Old Shep built a roaring pitch fire in the chuck wagon stove and cooked a big breakfast of steak, eggs, black coffee, and biscuits. Once I was mounted, Whingding lost his temper and tried to buck me off in the deep snow, but every time he put his head down, his nostrils plugged up, and he quit trying. Soon he forgot his outlaw ways as we went back to hustling the cattle, trying to gain the spot where the old road plunges over the edge of the rimrocks, where we could hope for a few miles respite from the blizzard.

Once, as the wind died for a time and the snow turned to sleet, we saw Morgan far up ahead, pushing a small bunch of lead cows so the others might have a track to follow. Midway in the herd, we could glimpse Jim O'Connor, bundled up to his eyeballs, flat-siding the flanks. We were so shorthanded Jim had to work both sides, tucking strays back into the herd, crossing over from left to right or right to left, whenever he saw a need. The snow had frozen to his clothes and the right side of his horse so that horse and rider seemed to have been sawed down the middle.

The cows still moved slowly, heads down, seeking out spears of last summer's grasses that managed to show above the snow. The older cows had used up what little strength they had left, and every mile or so we would find one dead or dying on the trail. There was not much we could do except take comfort that soon the fallen animals would have a proper burial beneath a soft blanket of snow.

It was late that afternoon when we drove the cattle into a holding field at the edge of the Sycan River. Ern Morgan had bought a stack of wild hay from Bart Shelley, who lived at the Sycan Bridge. We had no way to handle the hay and scatter it, so we opened the gates and let in the whole herd of hungry cattle to eat. We tied our horses in an old barn out of the storm, and went into Bart's house to thaw out.

There were two main rooms, but one was filled clear to the ceiling with old western magazines which Bart had read and discarded. I could hardly walk without stepping on a cat, and Bart had a habit of spitting tobacco juice on your boots as he talked. He wanted us to put our bedrolls on the floor, but we muttered something about checking on the cows, and soon had our tent set up out in the snow.

That night the wind howled and shook the canvas as we tried to sleep, but soon the snows drifted over the tent and all was silent. Shep had slept in the chuck wagon, and in the morning he dug the snow from in front of the tent and rousted us out. We could smell bacon frying and the fragrance of coffee and dressed in a hurry in the cold.

We knew the town of Beatty lay somewhere south of us, but the blizzard still raged and familiar landmarks were few. Our best chance to save the herd was to cross the Sycan at Bart's Bridge and head across the flats toward Paiute Camp and the pine forests north of Charley Mountain and Five Mile Creek.

We had gone perhaps three miles when the drag slowed

and stopped, and I trotted back to see if the kids were having trouble. The two boys were nowhere to be seen. From their tracks, they had left the herd and set off to try to find Beatty, abandoning us to the storm. I took over the drag myself, riding back and forth in the blinding snows, pushing cows as best I could. Minutes later, Jim materialized in the whiteout. He took in the situation without comment and went to work pushing animals across the sagebrush plain.

"Those kids," he mused. "Be the Jaisus, those kids are sure going to miss out on a lot of fun!"

Three of us, six hundred cows, and one of the worst storms in history. In the good old days, in good weather, we used to figure one cowboy for a hundred cows. The lives of every one of those cows depended now upon three men and an old chore man. There wasn't a choice but to carry on.

At noon Morgan appeared through the drifts, so caked with snow that at first we did not make him out. He saw at a glance that the kids were gone and Jim and I were pushing the herd alone. But there was more on his mind than that.

"I missed the trail in the storm," he said. "Instead of heading east and to the north of Charley Mountain, the cattle took off south until they hit the railroad tracks and turned east along the tracks at a long trot. No way I could hold 'em. They got the Sprague River on their right and rocky cliffs on the other. If the trains are running, the cattle will have no way to escape being killed."

Morgan rode close to us so he could read our faces. "Listen good," he said. "If a train does come along while you're on the track, you won't have a chance to save yourselves or the cattle. I'm going to try to push the cattle through, but I'm telling you both to head on north of the mountain. You got that? I'm ordering you not to come with me."

As Morgan trotted back to talk to Shep and send him north with us, Jim and I grinned at each other and moved off after the cattle. When Morgan caught up, we were already hurrying the cattle down the narrow track as fast as we could go.

As luck would have it, we moved the herd through the steep cuts of the railroad grade without getting caught by a train. We left the track at the Elder place, hoping we would catch someone at home. The hungry cattle milled around the buildings looking for shreds of hay. The Elder cattle had been trailed out to the desert earlier in the fall, and there were only a few butts of stacks left in the stackyards.

Morgan slipped from his horse and pounded on the door of the house. Snow cascaded from the tin roof, burying the old cowboy in the pile. There was a screech of rusty hinges, and the door opened a crack. One of the Elder boys stood looking out in surprise at Morgan buried to the waist, then at the hungry cattle milling around his yard. But he was quick to gather his wits. "Come in!" he said. "There's hay in the barn for the horses, and the cattle can make do in the stackyards!"

None of us had a word to say, for at that moment, a logging train came thundering past, its snowplow flinging up clouds of frozen snow.

The next morning we left the Elder place in a storm that seemed to have doubled its fury. Ernest Paddock had sent out a four-horse team with a wagonload of hay from the BK, six miles away, and the team had broken a trail through the drifts. I led Jim's horse, and Jim rode the hay wagon, tossing flakes of hay from the wagon racks. The cattle herd flowed on behind the wagon, their backs white with new snow. To our right, whenever the storm moderated, we could catch glimpses of the great Bly Valley, and occasional distant ranch buildings huddled against the storm.

Along the county road were miles of split-rail fences with only a few top rails showing above the drifts. The cattle seemed to sense that they were nearing the end of the trail, for they ambled down the road, heads hanging low, too tired to try the drifts on either side. At the top of a rise, the ranch tractor had cleared the drifts from a gate, and here Morgan turned the point of the herd off the county road. The cows gathered speed, and soon the whole herd was trotting down the hill toward a level field where Paddock's men were using pitchforks to roll hay off loaded wagons.

That night we sat at the long ranch table in the BK dining room, thankful to be warm once more. We would rest up for two or three days, take fresh horses, and set out a-horseback over the long road back to the ranch. With luck we would make it with only one night spent on the trail.

Chapter Eight

IT WASN'T LONG AFTER THAT MURDEROUS CATTLE DRIVE in the snow that the Yamsi crew left me to batch it alone. One by one they vanished from my life. Once she thawed out, Margaret Biddle went south to a retirement community in Santa Barbara. Buck stayed warm in a geothermally heated apartment in Klamath Falls or made the rounds of health resorts in California. Old Shep could generally be found in Main Street coffeehouses; Jim O'Connor went to live with a daughter; and Ern Morgan abandoned lifetime work in the cattle industry for wartime work in, of all places, a California doll factory owned by his new wife. It left me boss by default, and I could have hired and dismissed at will, except there wasn't anyone left to fire.

Not unless you counted Al Shadley. Al was drunk as a boiled owl when I first saw him. I had cut up a dozen big ponderosa pine logs into blocks with an old Wade drag saw and was resting on the pile in the snow, getting my strength back, contemplating splitting the blocks into my winter wood supply, when an old thirty-eight Chevy pickup came thundering down the road from town and made a wild, skidding turn into the ranch gate.

All four tires were wrapped with chains for traction, and the front bumper was pushing snow. A left rear tire was flat, putting out clouds of smoke. Only the snow kept it from catching fire. In the back of the truck was a spare tire, also in complete ruin, an old, battered snow shovel, and a rusty five-gallon can of swamp water filled with shiner minnows, most of which had sloshed over the lip of the can and were dead and frozen on the floor. Sticking out over the tailgate was a rusty spinning rod, with a big ruby-colored glass tip that was worn through to the point I got the notion the owner fished a lot more than he worked.

The jack pine sapling wedged in the well of one front tire was packing in the snow, making the truck difficult to steer, and the old man accounted for two more small trees before the pickup belched steam, farted, and lurched to a stop just inches short of my shins.

The old Indian rolled down his window and sat for a few moments looking at me, silent but grinning. He swung his head back toward the bucket of minnows and the rod. "I'm goin' fishin', see. I'm goin' fishin' on your propity. I fought like hell through thirty miles of driftin' snow to get here, so I won't take no for an answer. On the other hand, I don't want to end up in jail, so I just came by to let you know I'm goin' fishin' on your river, see."

"Sorry," I said, moving back from his breath. "The cattle rustlers have been working us over lately something fierce, and the only folks we let on the property are guys who work here."

The old man looked at me with a bemused smile as

though he well knew what had happened to some of our beef. He held out his gloved hand.

"This old Pit River Indian has a real hankerin' to fish that river," he said, grinning, "so shake hands with yore new hired man."

He shoved open the door of his pickup and fell out face-first into the snow, pulling himself up spitting, sputtering, and blinking through thick, iced-up lenses to where he could lean on a battered fender. There was a graveyard of dried insects left over from summer on the radiator that looked like the collection from a fisherman's hat, and he brushed them off with a wave of a gloved hand. Staggering to the huge pile of wood, he picked up a splitting maul and began to split the drums.

Suddenly he looked toward me where I stood, a little astonished at how well he had controlled the conversation. "I know I don't look like much," he said with a sly grin, "but believe me, yore gettin' one helluva man!"

I went off a-horseback through the snow looking for strays and came back that evening expecting to find Al gone or passed out beside that huge mountain of blocks without much done. Instead, my winter wood supply was neatly split and piled, and from the window of the kitchen came the fragrance of frying venison. On the table as I entered was a square cast-iron frying pan of fresh, hot frying-pan bread. He was, as promised, one helluva man.

We moved in three hundred brood cows from the Hoyt Ranch north of us, and fed them that winter by team and wagon, which meant handling lots of hay by pitchfork, seven days a week, but Al never complained. He turned out to be one of the best men with a team of horses I ever saw.

The big man was a Pit River Indian from northern California, and we were surrounded by a reservation which was mainly Klamaths and Modocs, with a scattering of Paiutes and Yahooskin Snakes. When a local Indian died, the relatives were in the habit of locating Al and his brother, Amos, giving them a fifth of whiskey to dig the grave.

Al hadn't touched a drop of liquor since the day I hired him, but one day, when we were both starting to get cabin fever pretty bad, he came to me and said, "They're havin' a big Modoc funeral tomorror over to Beatty an' they want me an' my brother, Amos, to dig the grave. Okay?"

"Nokay!" I snapped. "I've heard tell about you and Amos. You'll start partying, and it will be a week before you get back. There's no way I can feed all these cattle by myself."

"Well, maybe I should quit this goddam outfit then," Al said, his feelings hurt.

Pretty soon I heard his pickup truck start, roar skidding and sliding out of a snowdrift, and go popping up the road toward town.

They partied all night, Al and Amos, and things got pretty drunk, but the next morning, give them credit, they were out there at dawn in the cold at the Beatty Cemetery in Al's old truck, getting ready to dig the grave for a two o'clock funeral.

But they had forgotten one important thing: despite the blanket of snow, the ground was frozen hard and deep. Two hours later all they had was a hole down in the ground only a small rabbit could have used.

"Amos," Al said. "I think we better take off for Klamath Falls and get some dynamite. I know the old sheriff in there pretty damn good. Hell, he'll give us a permit for a whole pickup load of the stuff. How much money you got on you?"

They went to town and came back with the explosives.

"Amos, we better hurry," Al said, getting out of the truck. "Those people goin' to be comin' out from Beatty soon, an' we better have this hole ready for that pine box."

But it was cold, and first they had to sit in Al's old truck listening to some country-western music while they warmed their fingers and raided Al's battered old thermos for coffee. Gingerly, they stuffed the rabbit hole with all the dynamite it would hold, and hooked the pickup to a big snow-covered rock with a chain. The rock tore the bumper half off before it decided to move. They skidded it to the gravesite and slid it on top of the hole.

"That rock ought to hold down the charge," Amos said as he unhooked the chain.

"Looks pretty good to me, Amos," Al said. "I think we can touch her off now and build our hole. What kind of whiskey you s'pose they'll give us anyway?"

"Hey, look at that, would you?" Amos said, squinting down the road toward town. "I can see them comin' already. Cripes, I didn't know there was that many cars on the reservation."

Under his buckskin gloves, Al's fingers were hurting

with cold. "Damn it to hell, Amos," Al said. "It's coldern a well digger's ass out here. I'm goin' over to the pickup and have me some more of that coffee."

"Hell, we ain't got time," Amos said. "We gotta touch this thing off."

He stepped back sharply as Al dropped a stick of dynamite on the frozen ground. "You be careful," Amos warned. "All these people sleepin' here underground, you want to wake 'em up?"

In the distance, a whole line of cars proceeded to turn off the main road into the cemetery.

Al stuck a shiny metal cap on a piece of long black fuse, crimped it with his teeth, poked a hole in the dynamite with a sharp stick, inserted the cap, then stuck it as a detonator into the explosives packed beneath the rock. He split the other end of the fuse with his pocketknife to expose the black powder and make it easier to light.

"Lookit all them people," Amos said. "Maybe we ought to motion 'em to move back. I never seen so many people. Either these Modocs like the guy that's bein' planted or they hate him and want to make damn sure he's underground!" Amos pulled out a hip flask he had been hiding under his coat, took a big swig, and handed it to Al. "Here, brother, you take some of this now. It'll warm you up quicker than gettin' laid."

The music from their pickup radio seemed a little subdued by now, as though the battery were on its last legs. Al lit the fuse, and as the sparks flew out over the frozen earth, a pall of white smoke spread out like a deadly gas. The two men scrambled to get behind the vehicle for safety.

As the charge went off, the lead car of the funeral procession slammed to a stop. Traveling far and wide under the frozen ground, the dynamite shook the earth; the whole cemetery seemed to leap into the air in a cloud of white smoke.

"Well, look at that, would you?" Amos hollered above the din. "We just blew up the whole damn graveyard! We better get out of here, Albert. Pretty soon those Modocs are goin' to start collectin' some Pit River scalps."

A large coffin fell out of the air, narrowly missing the pickup, and spilled its contents over the ground.

"Hey, Amos!" Al exclaimed. "Damn if that ain't old One-eyed Alice. I used to spark that woman. Damn if she don't look better now than she did then!"

The two men jumped into Al's pickup to make their getaway, but the radio had been too long a-playing. The starter clicked once and stopped. The battery was dead. Abandoning the truck, they ran for their lives and lit out for the ranch afoot, twenty snowy miles away.

Their last peek at the graveyard showed total destruction. Trees uprooted, coffins standing on end, bones, skulls, headstones, and plastic flowers all in one great jumble together. It was the Fourth of July before Al dared to go to town again.

Chapter Nine

THAT SUMMER, ROSE WAS CHOSEN QUEEN of a big Oregon rodeo, and the newspapers had a lot to say about how well she rode a horse and how smart she was in school. She had beaten out several other girls for the rodeo honor, and I knew there were some hard feelings, especially among parents of some of the horsey white girls. For the Indian community, her winning was a source of great pride. They tried their best to make Rose's victory their victory, and offered to loan her fancy beaded riding outfits that would not have been out of place in museums of Indian art.

Rose, of course, would have none of the fancy stuff. She rode in faded Levi's, with a cotton shirt she had made herself, and when she galloped past the rodeo stands, the wind pressed the thin material against her slender young body and brought a gasp from the crowd. It was less a tribute to femininity than a triumph of grace. She rode like the desert wind in the sage.

The Indians had some fancy horseflesh on the Klamath and Warm Springs reservations, and Rose could have taken her pick of queenly mounts, but she stuck to her old paint horse and made that half-mustang pony seem elegant.

I was proud of her, of course, proud of what she had been able to do for herself, proud that of all those hundreds of people in the crowds I was maybe her best friend. That evening at the rodeo dance, she was thronged by admirers, but she left them all to find me in the crowd and pull me out to dance with her. That started a few unwarranted rumors, I'm sure.

Over her shoulder, as we danced, I kept getting angry stares from some of her uncles as well as boyfriends, and I couldn't shake the feeling that I'd lost some Indian friends valuable to me and from now on maybe I'd better watch my back. There were still those Indian people around who clung to the old language and considered other races trash. For being rodeo queen Rose had been given a small scholarship to further her education. She had asked me to dance for one simple reason, to say good-bye.

"I'm going away this fall," she said. "I've always dreamed of being a nurse, and with my scholarship money and with what I've been able to save by working, I'm going to nursing school in San Francisco."

I was stunned, of course. Down deep I knew I should be delighted by her plans rather than depressed by my loss of her friendship. "I think it's great," I told her and concentrated on dancing with her for what might be the last time.

"And what about you?" she asked, looking up at me.

"I don't know," I admitted. "There's the ranch, of course, and I'd like to rodeo some. And then there's the war. I could maybe get an agricultural deferment, but it doesn't seem right that other guys are out there dying. I suppose I've already made up my mind to go into the service, but I'm afraid to face my uncle."

We danced another couple of rounds in silence. She wasn't her usual self, and I suspected it had nothing to do with being rodeo queen.

"What's wrong?" I asked. "This ought to be a great time for you, but you don't seem very happy."

"I guess I'm scared." she said, dancing away from the other couples so no one heard. "I'm having a bad time with my people. I guess they're jealous of me, afraid I'll succeed and leave the Res." Her face showed sudden anger. "They sit around the reservation on their asses, blaming the whites for all their troubles, afraid one of us will succeed and show the world it can be done. My brothers drink and give me hell when I won't drink with them. I guess I'm a social creature. It hurts like hell when my friends and relatives give me the silent treatment."

I was puzzled by her outburst, feeling somehow that I had to defend my Indian friends. "Some Indians here have done pretty well," I said. "Look at Orrie Summers, Mamie Farnsworth, and Dally Givons, for instance. They are all respected livestock people."

"You look at them," she said sadly. "Every one of them came from somewhere else and left their Indian baggage behind."

"Then that's what you have to do," I said. "Go to nursing school and to hell with your family."

Rose looked a little brighter for a moment, then the old fears returned. "I'm scared," she said. "Scared San Francisco isn't far enough away for them to leave me alone."

One of her uncles cut in on us then and herded Rose away. I saw her a few times after that, but always in the distance. I often asked some of my Indian friends about her, but they had a frozen-faced way of giving me a nonanswer. I knew that I wasn't going to help Rose by butting into her life. She was wise enough to know what she had to do, and she would have to do it alone.

Chapter Ten

MY ONLY SOURCE OF NEWS AT YAMSI during the winter of 1943–44 was an old battery-operated Philco radio, which, in a way that seemed to me to be miraculous, brought in news over the mountain passes which were buried in ten feet of snow. But by February the battery had gone dead, and for all I knew the war could be lost or won on the part of America. I began to harbor a real guilt that I hadn't yet contributed and that, locked in behind my mountains, I was probably the safest man in America.

Now and then I would get a V-Mail from North Africa or Italy from a favorite BarY cowboy, Tommy Jackson, and I'd feel a twinge of guilt that I wasn't doing my part. At night, I lay in my bed running through endless dialogues with my uncle telling him that I had decided to enlist. Always I envisioned the old man coming up with one big question: "Who the hell is going to take care of the ranch?" Snowmelt came, and when my uncle returned from California, he started right in regaling me with his troubles, and I ended up afraid to tell him. He'd lost cattle to rustlers that winter, and due to the war there wasn't much left in the way of law enforcement to help him.

Gas rationing had limited my ability to patrol, but the next morning I rode BK Heavy seventeen miles to the Calimus Field to check some yearlings, then rode on west cross-country, searching for Joe Perry's cabin to try to get the old man to come work at Yamsi. I needed someone to keep the fires burning when I was off with the cattle.

I was approaching a meadow called Long Prairie when I noticed a flour sack, neatly tied with baling twine, shoved under a fallen log in the forest. In that sack was the hide of a BarY steer freshly butchered by rustlers.

Rustling was a penitentiary offense in that country, and in any given year there was often a convict up in the Oregon State Pen doing time for stealing my uncle's cattle. We generally knew just which of the locals were stealing BarYs, but making a case was difficult. Underneath his solemn exterior, Buck had a kind heart, and sometimes he was his own judge and jury. If the man butchered a BarY to feed his hungry kids, my uncle was likely to look the other way. After a fashion, he was living the Cree Indian cannibalism case all over.

One of the neighbors made a game of it, and when he would pass Buck on the street he would slap him on the back and exclaim, "Come on over to eat with us sometime, Buck. The old lady will cook up some of your beef!"

Buck owned some of the best Hereford bulls in the country, and sometimes another neighbor, Jim, would stop him on the street and say, "Couple of years ago I stole two of your bulls. I'm done with 'em now and don't want to feed 'em over the winter. What do you want me to do?"

Buck would send us over a-horseback to get the bulls,

and it was sort of understood that Jim would help himself to another couple the next spring. In fact he used my uncle's bulls so long, his cattle began to look like BarYs.

I was pretty angry about finding the butchered BarY steer, however, and determined to find out who had butchered it. I had just ridden a mile or two down to the road along Long Prairie when Jim's son, Albert, came riding by a-horseback.

"Albert," I said by way of greeting, "who cooks with Pillsbury's Best flour in this country?"

"Why, Dad cooks with Pillsbury's Best," Albert said.

"Who bales hay with New Holland twine?" I asked.

"Why, Dad bales hay with New Holland twine!" Albert paused a minute thoughtfully. "My gawd!" he said. "Do you suppose Dad butchered that BarY steer?"

I'd ridden with Albert since I was a kid, and he was a favorite friend. Now I could just grin and ride off on my way, shaking my head. I suppose that sack with the steer hide is there to this day.

Joe Perry (actually his name was spelled Parais, but everyone knew him as Perry) was out in front of his cabin, feeding birds, when I rode up. Joe was a little bit flighty, and I never could tell whether he'd meet me with a rifle or a smile. A Louisiana Cajun, he'd done time in Leavenworth, where, it was rumored, he'd killed a guard. He'd ended up on the reservation, living in a small cabin on Long Prairie and generally staying out of trouble. His main difficulty, as I saw it, was that he was a great storyteller and everyone liked him. Certain women, perhaps sensing the magnetic force of his ready availability, wouldn't let him alone.

I rode old Heavy through his gate, and as I got off my horse, Joe made about three tours around him, spitting out tobacco juice onto the bitterbrush as he inspected the animal.

"Reminds me of a horse I had when I first come to Klamath," Joe said. "Man, could he trot!"

BK Heavy wasn't too comfortable to be inspected by a man packing a rifle, but after a loud snort or two which scared the chickadees off the bird feeder, the horse settled down, grabbed a mouthful of bitterbrush, and commenced to chew while Joe sat down on a stump.

"One winter day," Joe said, "one winter day I was ridin' thet ol' hoss down Long Prairie Road when Frank Summers come along in his pickup and stopped.

" 'Joe,' he says. 'It must be coldern hell out there. Git off yore hoss and climb in this cab and get warm. You can roll down the window a little and lead that old pony by the reins.'

"We had gone along like that about a mile and had dropped the level of Frank's whiskey bottle 'bout three inches when Frank said, 'You know, from the looks of thet old hoss of yourn I bet he could really trot.'

"I says, 'Yessir, headin' down the road fer home, he can really fly.' We had another drink of whiskey, Frank an' me, and got to timin' the old hoss on the speedometer. Frank got to goin' faster an' faster an' thet old horse got to fairly flyin'. I looked out the window an' says to Frank, 'I think maybe you ought to slow down a little, pardner. Why, thet ol' hoss's feet ain't touched the ground in a mile an' 'is head is stretched out six feet long!' "

Joe might not have been too enthused about helping me on the ranch, but by the time I got back to Yamsi there was smoke coming out of the bunkhouse chimney and a big greasy pot of porcupine meat stewing on the stove.

"Reminds me of when I first come to this country," Joe said, stirring the pot. "I eat a lot of porcupine 'cause they is easy to catch in the woods."

Joe put another log into the stove, and we both had a fit of coughing as a sudden backdraft filled the room with a layer of smoke.

"How did you happen to end up in this country, Joe?" I asked, hoping for a story.

"I come here through California an' was tryin' to catch a freight from Sacramento to Chiloquin. Well, I got on a freight goin' north all right, but by the time we passed Chiloquin, we were goin' a hunnert miles an hour, an' I ended up in Portland. Well, I caught a train headin' south an' when we passed through Chiloquin, dam' ef we warn't goin' a hunnert miles an hour again an' I ended back in Sacramento.

"Finally," Joe said, "I got tired of travelin'. Here I was goin' south again from Portland, an' we hit Klamath Marsh goin' a hunnert miles an hour. Well, I jumped off thet train and landed in pummy dust right up to my chin!

"Hadn't been in Chiloquin long when I married the biggest lady on the reservation, but she died."

"I'm sorry," I said, aware I was about to be had. "What happened?"

"You might say she died of bad eyesight," Joe said. "She

was wadin' in the Williamson River one day without no clothes an' she looked down at her reflection in that clear water, thought she saw a canoe, stepped in it, an' drowned."

I was about to leave the bunkhouse, afraid I would have to eat some porcupine, when Joe spotted a photograph someone had tacked to the bunkhouse wall. "I know thet woman in thet picture," Joe said. "I was drinkin' one night in Chiloquin when she come up to me an' sez, 'Joe. My old man an' I need a ride out to Sprague River. You live out thet way. How 'bout you give us a ride?' "

Sprague River was only a few miles out of his way, so Joe agreed. They had a few drinks for the road, and Joe loaded the woman and her husband into his pickup. It was winter and below zero, but the ice on the windshield soon melted from body heat. The lady took up most of the front seat, with Joe and her husband crowded up against opposite doors.

"We passed my bottle around a few times," Joe said. "When I looked over at her husband, he had passed out. I said to that woman, 'Now that was an unmannerly thing for him to do,' so I pulled over to the borrow pit, reached across her lap, opened the door, and let him fall out into the snow.

"We were goin' by the Lone Pine turnoff and the lady said, 'Joe, I've never seen your cabin. I want to see where you live.'

"Well, we stopped at the cabin an' had a few more drinks. I left her sitting on the edge of my bed and went to put another log on the stove. When I came back, she had passed out cold.

"I thought, Now that was an unmannerly thing for her

to do, so I went out to my shed, got a big bucket of green paint, lifted up her dress, and painted her green."

Joe had agreed to come out to the ranch for a few days and help me, but he said he had to be back in a couple of weeks to go to court. He had put out some seed for his chickadees and some apples for the porcupine that lived in a pine tree beside his cabin, loaded some venison and a bedroll into his pickup, and had been ready to go. Even then I was surprised that he actually made it to the ranch without changing his mind.

"It's none of my business, Joe," I said as we ate supper at the ranch, "but what's that court date all about?"

"Oh, I just killed a guy," Joe said. "I was bangin' on this old lady in my bed when someone started poundin' on my door. It kind of scared me an' I thought, Now that's an unmannerly thing to do! So I took my old rifle and shot through the door, and there was no more of that noise. When I went out later, there was the lady's husband layin' there dead with a bullet hole in him. Hell, it was all an accident. I was just tryin' to scare him away."

The chickadees around the ranch seemed to like Joe, for they called to him from the pines, and one flew down and sat on his arm. I rode down through the ranch a-horseback, knowing that Joe was a man of his word. He would stay at the ranch until his court date and be ready and willing to help.

We got along fine, Joe and I, but you can bet I watched myself so as not to do an unmannerly thing.

Dayton O. "Hawk" Hyde at sixteen with his uncle, Dayton O. Williams. *(Margaret Biddle)*

ABOVE: Ed Donovan riding Blackhawk at Beatty, Oregon, 1940.
BELOW: Dayton Hyde winning saddle bronc riding at Marysville, California, in 1944.
(Photo, U.S. Army)

TOP: Hyde's trip from Louisiana to Oregon with plantation walking horse mare Sugarfoot, who foaled many Hyde colts.
ABOVE: Dayton Hyde as rodeo clown, fighting bulls at Hawthorn, Nevada, 1949.
(Photo by William D. Schnack, taken with Hyde's camera)

ABOVE: Dayton Hyde luring the bull away from cowboy Lucky Buck, Hawthorn, Nevada, 1949. *(Photo by William D. Schnack)*
BELOW: Slim Pickens caping a bull in California rodeo.

Slim Pickens in Billings, Montana, 1948. Slim was an all-around rodeo athlete, bullfighter, and clown before he became an actor. *(Photo for* Life *magazine by Dayton Hyde)*

Slim Pickens riding his horse Honest John in Yamsi corrals. John, who was trained to buck on cue, starred in *The Big Country,* bucking off Gregory Peck.

Famous rodeo cowboys, *left to right:* Casey Tibbs, Jack Sherman, Buster Ivory, Bill Linderman, Chuck Shepard, and Stub Bartlemay.

TOP: Rodeo clown and all-around cowboy Felix Cooper got his start in Florida, taking grapefruit off the horns of fighting bulls for pennies.
ABOVE: Clowns Slim Pickens and Felix Cooper *(right)*.

Lloyd Lippi comes off Harry Rowell's Bull #13. He was uninjured. This became a *Life* Picture of the Week, and the bull was thereafter named Life Magazine.

Hyde took this photograph for *Life* magazine's "Speaking of Pictures," November 1948.

Cecil Henley, a fine old California bronc rider.

Bud Linderman on a saddle bronc.

Oregon cowboy Sonny Tureman on bareback bronc at Red Bluff, California.

Dick Stevenson, Dorris, California. Dick was killed soon after in a light plane crash. This photo won a Grafley Award and a place among *Salute* magazine's Best Pictures of the Year.

LEFT: Jerry Ambler, world champion and a great balance rider, riding the bronc Red Bluff at Red Bluff, California.
RIGHT: Jack Sherman in Pendleton Round-up finals for saddle bronc riding, Beatty, Oregon.
BELOW: Ross Dollarhide comes off Mac Barbour's Warpaint at Pendleton.

TOP: Ross Dollarhide showing championship form on a tough Rowell bronc in 1948.
ABOVE: Many-times world champion Casey Tibbs on a bareback bronc. Tibbs was one of the great balance riders.

ABOVE: The great Joe Kelsey bronc Snake at the Pendleton Round-up, Beatty, Oregon.
BELOW: Ross Dollarhide in 1949.

Boyd Hicks coming off the famous Rowell bronc Sceneshifter. Harry Rowell watches from the chutes.

Dayton Hyde and his friend Mary Jo Estep, last survivor of the massacre of 1911 in northern Nevada.

(Photographer unknown)

Opening ceremonies at Black Hills Wild Horse Sanctuary, 1988, *left to right:* Mel Lambert, Montie Montana, Dayton Hyde, Betty Zane Breslau, Governor Michelson of South Dakota, and Robert Burford, director of the Bureau of Land Management.

Hyde driving his team of Percherons, Pat and Mike, in the movie *Crazy Horse*.
(Photo by Jim Hustead)

When Willie came off a BLM truck at the sanctuary in 1988, she was too weak to stand. Hyde lifted her bodily, leaned her against the corral fence, and coaxed her to eat and drink. Seventeen years later, she still remembers the weeks of loving care, and gallops up at the sight of him.

Chapter Eleven

WITH WORLD WAR II IN FULL SWING, and not much gasoline available, I spent more and more time a-horseback. Occasionally the employment office in Klamath Falls would send out a warm body in response to my pleas for help, but no one seemed to stay very long. Most of them came out from town bleary-eyed and smelling of cheap wine, seeking a place to dry out and reform, but after a week or two there was no holding them back from their thirst, and some of them simply disappeared, hiking the thirty miles to Chiloquin rather than face me.

Buck cut down his operation as best he could until the only regulars left were Ernest Paddock at the BK Ranch at Bly and myself at Yamsi. There was not much communication between Ernest and me. Being the owner's nephew was not the easiest cross to bear, and as yet, in spite of trying hard, I had not proven myself a very handy cowboy. One trouble was that when I began to make some pretty fair shots with a rawhide reata or rode a bucking horse, there was no one there to see.

I was taken by surprise one day when I came into the ranch kitchen with an armload of wood, to hear the old crank telephone ringing away. It was Paddock on the line, calling from the BK. I expect a wild party was in progress, for there was lots of drunken laughter in the background. Ernie was actually very friendly, a fact that made me instantly suspicious. And furthermore, he wanted to make me a present of a horse. Not any old horse, he said, but one to drool over.

"I've got a King Ranch quarter horse from Texas," he said. "He's the prettiest little bay gelding you ever saw, and I'm in the notion of giving him to you. Throw your rigging in the pickup and come over here right away!"

Paddock knew his horses and wasn't about to get rid of an expensive quarter horse from the King Ranch in Texas unless something was very wrong with the animal. Since the pickup gas tank was dry, I had no transportation beyond my horses and was about to tell him I wasn't interested, but somehow the situation intrigued me. A few minutes later I had dropped my saddle on old Whingding, and was on my way through the back country to Bly.

By the time I had tied my horse in the BK barn the next day and run the gamut of a dozen back-slapping drunks, I knew that there was more to Paddock's generous gift than met the eye. I had a sinking feeling that Paddock was giving a big three-day party, and that I had been set up as entertainment. When I walked down to the corrals to inspect the horse, a whole crowd found an excuse to go down the hill in step with me.

The horse seemed gentle enough. He was one of the most gorgeous pieces of horseflesh I'd ever seen, and I was sure Paddock had paid a lot of money for him. King Ranch horses were famous and cost a pretty penny. I caught up the beast and became more pleased and excited by the moment. Maybe the old foreman had finally recognized my true worth and decided to be nice to me.

I saddled up the horse, checked my rigging, ran him around the corral a few times to warm his back, hung a snaffle bit in his mouth, mounted carefully, and rode the animal around the corral. There was a murmur of disappointment in the crowd when nothing untoward happened. The horse responded well as I circled the round corral and did figure eights, then I nodded to Paddock to open the corral gates and turn us loose into the big field.

I had just ridden the colt up on an irrigation canal bank when he figured he had me, dropped his nose between his forefeet, shot skyward, and came down bucking hard. I lost one stirrup, got it back, then lost the other. Every jump popped me up a little higher above the saddle, and I was starting to look for a soft place to land. Out in the middle of the field, the horse gave one big freak jump that should have sent me flying, but by accident I came down hard on the saddle. There was one split second when I considered jumping off, but I doubled the animal left and right, then rode him calmly back to the corrals.

Coming through the gate, I rode straight up to Paddock, who was standing there red-faced and flustered, angry that he had given that horse away. "You got any more horses around you old bastards can't ride?" I asked.

It turned out he had quite a few and was tickled to death that I asked. That afternoon, I saddled up one nasty horse after another and rode them until they quit. The last one was a big bay horse that ended up on top of the hayrack with me, and it took the rest of the afternoon to extract the animal and get him back to earth.

That afternoon gave me a confidence I'd never had before. There were some good old ranchers in that crowd of spectators, and I suddenly realized I was going to have to try my hand in the rodeos someday and determine just how well I could ride a bronc in competition.

It was not a good day for Ernest Paddock. He had just bought a beautiful set of draft harness for a team of Percherons he liked to drive in parades. He had hung the harness in a special place in the barn where he could show it off to visitors. We were having supper that night in the BK dining room when there was a knock on the door and in came a neighbor named Lester Hixon. Old Hick had a set of harness he wanted to sell Paddock, and his description of that harness was enough to make a teamster reach for his wallet. When Lester dragged the harness in across the floor, the leather gleamed and the silver conchas lit the room with a thousand little moons. Paddock was hooked and gave Hixon a pot full of cash and a fifth of whiskey for the set. "It's lots prettier even than the set I bought the other day," he gloated as Hixon left. "And did I ever get it cheap!"

After supper we helped the old foreman pack the heavy harness down to the barn. It was an awful moment for Paddock. The harness hooks were empty. Hixon had sold Paddock his own rigging.

Not long after Lester Hixon sold Paddock the harness, he was arrested by livestock theft officers for possession of about fifty head of horses he had stolen from his neighbors. The law impounded the horses over in Medford, Oregon, some two hundred miles away, and held them as evidence in Hixon's coming trial. Hixon himself was released on bail.

The case never came to trial, for someone crept into the Medford stockyards one night and stole the herd of horses. The tracks led east to the Oregon desert, where the animals scattered like the winds and were never seen again.

Hixon was eventually sent to the pen on other charges. There he distinguished himself by doing leatherwork and silversmithing, creating items which were of great demand amongst his former neighbors. He sent Jack Morgan a breast collar inlaid with silver hearts, and Margaret Biddle acquired a half-breed bit whose shanks were inlaid with silver hearts, diamonds, spades, and clubs, all done by Hixon while he was doing time.

I was afraid that Paddock would renege on his promise to give me the bay horse, but next morning he was out in the corrals helping me halter the little bay and tie the lead rope to Whingding's tail.

In a week, I was back at the BK a-horseback to pick up a string of colts to start, for Yamsi was in dire need of horseflesh, even though it was hard to come by cowboys willing to ride. I left the BK early one morning leading five colts tied nose to tail. I rode north of Bly past the Obenchain Ranch and headed for the Sycan Marsh and the ZX, where I would spend a few days.

The ZX was then owned by Kern County Land and Cattle Company down in Bakersfield, California, and butted up to my uncle's ranch. It was a big tradition-bound outfit with a million and a half acres of land, including the twenty-three-thousand-acre Sycan Marsh ranch. I rode down out of the hills to Sycan headquarters and was greeted by the sight of over a hundred fine registered broodmares that had all been bred to a jack and were nursing mule colts.

I fed my horses and found an empty cot in the bunkhouse. One by one cowboys kept drifting in and, tired from a long day's work, lay on their cots rolling Bull Durham cigarettes and waiting for supper. A few of them did leatherwork, mending tack and splicing reatas that had been broken in the day's work. We washed carefully in enameled basins filled from a hand pump and threw the gray wash water off the bunkhouse porch into the dust.

There were several beef carcasses hanging in a shed near the cookhouse, and I wandered over as the cook took a canvas off one of the beeves and commenced cutting it all up into great piles of steak. To cowboys everything was steak, and no other cuts were edible. There was no meat rationing here on the ZX, and the war seemed very far away.

Before daylight, we went out into the chill of morning, ate breakfast at the cookhouse, saddled our horses, topped them off, and rode out across the Sycan River to gather up a big herd of ZX cows and calves. These we herded up into a fence corner so that the buckaroo boss could work the herd for dry cows. Not trailed by suckling calves, these cows tended to be in better condition than wet cows nursing babies, and were slated for market.

My uncle's cows had all been dehorned as calves and

were uniform in color and body. The ZX cows had long sharp horns and were big, rangy cattle as wild as deer. It took good cowboys and fast horses to keep up with them, but we were a skilled crew even if most of us were too old, too young, or unfit for military service. For part of the morning the work went slowly but well. We sat quietly on our mounts, holding the herd until the boss came at us with a dry cow, and then we would let that designated animal slip by and hold the rest.

Midmorning a rider came out from headquarters, riding a fancy horse and outfit. It was one of the big bosses who had just arrived from Bakersfield on an inspection trip and wanted to help. The cowboss stuck the man out on the edge with the rest of us waddies to hold the herd and went back to cutting out drys.

To the big boss, however, cutting out the drys seemed like a lot more fun than holding the herd, and pretty soon he spotted a dry cow and rode into the herd to get it. The cows were nervous and melted away before his strange horse, but soon he got the dry cow started out of the bunch. At the edge of the herd, however, the cow refused to leave and ran back to the others.

Again and again, the man tried the cow, and each time the cow would make a fool of him. By now the herd was getting restless with all that activity, and we were having a hard time holding them. I could see that the cowboss was getting a little red in the face over this breach of cowboy etiquette. It was his job to cut the drys and no one else's. The herd was getting ready to explode when the cowboss rode over to the boss from Bakersfield, touched the brim of his hat politely, and said, "Mr. Smith, why don't you just hold on to that cow and let us boys drive the herd away."

A few months later, in an act of corporate fiddling, the management in California sent up a young man named Ray Stansfield to take over from the old boss, Buster Vaughn. Ray was short, handsome, a graduate of Cal Poly, and had a good understanding of the cattle industry. He was teamed up with Rich Bradbury, a big, tall beanpole of a man who had been a cowboss on the ZX since before I could remember.

In areas where heavy auto traffic passes through fences, ranchers put in auto gates or cattle guards, which are often made of railroad irons placed over a shallow pit. Cars can drive over the rails, but cattle hooves slip between the gaps in them. On dirt roads, the pits beneath the rails require frequent cleaning. I was digging out a cattle guard one day east of Yamsi when Rich and Ray came along, driving a ZX pickup. I had the road blocked and made them wait. They sat bemused, watching me make the dirt fly. I might have been running my uncle's outfit by then and was sweating hard, but they were bosses of a really big outfit and didn't do physical work. I tried to con them into giving me a little help, but they didn't even have a shovel in their pickup. It reminded me that on a small outfit, one did everything; on a big outfit, you were highly specialized.

Bradbury must have said a few good words about me to Ray, for he invited me over to his house in Paisley for dinner. He and his wife, Ocie, lived in a great big ZX house. I arrived for dinner and nearly froze to death in the dining room. All I remember was that the wind blew so hard through the house, it kept blowing out the lanterns, and there were whitecaps in my coffee cup.

The ZX was famous for having a bunch of cranky, spoiled old saddle horses that were a pain to ride. They could kick your overshoes off in the stirrup, buck you off miles from anywhere after a hard day's ride, and were just as dangerous to a cowboy on the ground as in the saddle. Paying extra to have those horses ridden didn't make much sense to Ray Stansfield, since there were too many good, honest, gentle horses to be had. The former boss, Buster Vaughn, was still there helping with the transition of power when Ray ordered that every sour old outlaw of a horse on the ranch be sold. He contacted Mac Barbour, a rodeo stock contractor from Klamath Falls, and arranged to sell the horses.

When Ray mentioned to Buster Vaughn that he had a deal with Mac Barbour, Buster looked at him and grinned. "That will be something to see," he said. "There's no way a man can do business with Mac Barbour without getting taken. Absolutely no way!"

Ray Stansfield bridled a little. "That remains to be seen," he said. "I appreciate the advice, but Bradbury and I will watch every move the man makes, and we'll weigh up the horses ourselves on ZX scales. How's the man going to cheat us?"

Buster Vaughn just leaned back in his chair and grinned.

The next day, Mac Barbour showed up with his stock trucks and a group of cowboys. Barbour was a little, short, bustling man with pink cheeks and a nervous laugh, who knew as much about bucking horses as any rodeo promoter in the country. He would sell out his string to someone like Gene Autry, then come up with horses just as good within months of the sale. He got some of his best horses by lending Indians on the reservation money, then collecting the debt in horses.

He climbed out of his fancy car and shook hands with everyone. To Mac no man was a stranger.

"Ray," he said to Stansfield. "Your boys are awfully busy, and I've brought my boys with me. They all have their saddles with them; if you'll just lend us some saddle horses, we'll do all the work and run those spoiled horses in for you. When we get them in, you and Bradbury run the scales. We'll weigh up, then I'll hand you over the cash, and we'll be off down the road."

Ray flashed a triumphant look at Buster Vaughn, who sat on the corral fence and watched the procedure without saying a word.

The Barbour cowboys were mostly kids off the reservation and a good bunch of hands. They borrowed some ZX horses, saddled them up, and soon had the bunch of outlaws to be sold trotting toward the corrals.

Bradbury and Stansfield stood at the gate, counted the horses into the corral, cut bunch after bunch onto the scale, weighed them carefully, and balanced the scale after each and every draft. Everything went smoothly. Barbour paid for the horses in cash, and pretty soon the horses had been loaded and the trucks disappeared down the road toward Lakeview.

"Well, sir." Stansfield grinned as he approached the old boss. "That whole operation went smooth as silk. Rich and I watched that little scoundrel like hawks and made sure he couldn't get away with a thing. I told you we were too smart for him!"

"Oh, everything went fine," Buster Vaughn said, "up to a point. While you were weighing up those horses so carefully, Barbour's men loaded up the saddle horses his men had borrowed, and by now they are safely in California." According to some, Mac Barbour sold those ZX outlaws in Los Angeles as gentle horses, and for a time, you could see dude riders flying clear up over the skyline of the city.

As for the ZX, with the selling of those horses, the operation succumbed to modern times. A few months later, they decided that work teams were obsolete. I bought their draft teams just to give them a home, and used them to feed cattle. Their old wooden-wheeled freight wagons, that had once freighted supplies from Lakeview to Paisley across the desert, I bought for twenty-five dollars each. The wagons were considered junk by the new managers, but they still stand on the ranch at Yamsi. To us they are a precious bit of history.

Chapter Twelve

If my uncle was proud of the work I did, or the way I handled horses, he was damned careful not to let it show. Sometimes he would be standing on the street corner on Main Street, Klamath Falls, talking to his friend and banker, Ernie Bubb, and would be bragging on me as I approached behind him. I would catch only the tail end of a compliment before he saw me and froze up. Likely he would restart the conversation with a statement about how worthless my generation was, and how I had rodeos on the brain and was bound to get crippled for life even while my father languished in a wheelchair, a millstone around my mother's neck.

It was not that my uncle hated rodeos. He just didn't want me involved in anything but ranching. He thought Jerry Ambler was great. In time, Buck came to visit the Montgomery Ward's saddle department with increasing frequency and talk to Jerry about balance riding, a style where a rider rode loose instead of gripping the swells of the saddle with his knees. It took balance, not strength.

The old man never got on a horse himself, but he decided that he'd have Jerry design a saddle that would make a balance rider out of me. I was pleased that he thought of me, but when my uncle came over to the corral one day and dumped the new saddle on the grass beside me, I had all I could do to keep from laughing. The saddle had a nineteen-inch tree, in other words measured nineteen inches between the swells in front and the cantle behind. When I sat in it and brought up my knees to grip the swells of the saddle, I missed the swells entirely. I should have had it out with the old man right then and there, but he jumped into his Chrysler and was gone in a cloud of dust.

I was breaking some seven-year-old colts at the time which were a little old to train easily. When I could, I kept them from bucking, since they were big enough and strong enough to put me on the ground. The task was tough enough without riding a saddle that I couldn't grip. But my uncle was the boss, and I put my saddle in the barn and dropped the new rigging on a big bay colt that was my favorite in the bunch.

The leather was stiff and new, and the colt took an instant dislike to the noise it made, but in time the horse settled down. After a half hour of riding in the round corral, I rode out onto the range, intent on putting some miles on the colt to tire him out. It was just beginning to snow as I started up the ridge toward Taylor Butte and the Sycan country.

Halfway up the hill, the colt suddenly went insane and blew up. He bucked under trees, over windfalls, getting better at bucking with every jump, squealing in anger every time he hit the ground. Sometimes I would get him stopped, but as soon as I urged him to walk he would start again. I was riding on balance because I had to; there was no way I could grip that plunging saddle with my knees.

Just then the colt smashed into a tree, and the branches smacked my face, temporarily blinding me. In a split second I was on the ground, and as the colt whirled to kick at me, he jerked the lead rope from my hands and galloped off through the trees.

The snow was falling fast and furious as I trailed him through the woods. Moments later the colt circled and came up from behind me, following his own tracks. I tried to stop him and grab my lead rope, but I missed and the animal was gone, nickering wildly to the horse he thought was ahead of him. Soon snow obscured the tracks entirely, and I walked back to the ranch rather than be trapped in the woods by darkness.

The colt had been born on the BK, and it was natural for a horse to head back to the range where he was born. I called Paddock at the BK and told him to leave gates open so that the horse could come in on the ranch, but the animal never showed. Years later, a logger found the bones of the horse and the remnants of that saddle at the base of a forty-foot cliff, over which the animal had plunged in the snowy darkness.

I dreaded facing my uncle. His first words to me were, "How do you like that new saddle?" I had to explain to him that the horse had bucked me off and neither the saddle nor the horse had been recovered. He might at least have asked me if I'd gotten hurt, but instead he went storming off to town. It wasn't hard, the next time I saw him, to tell him that I was joining the army.

I had hardly given him notice when we were scheduled to move several hundred steers from a pasture in Fort Klamath to the railroad stockyards in Chiloquin. To get to the yards we had to cross the tracks and had been given clearance by the stationmaster. We were delayed in crossing the bridge over the Williamson by some Indian boys who demanded ten dollars before they would get out of the way and let the animals pass. An unscheduled wartime freight train roared down from the north and smashed into the herd just as they were crossing the tracks, killing sixty of the animals. I felt sorry for the old man then.

Waiting to hear from the army, I helped ship some four thousand cows to California and planned to drive them through the town of Williams west into the hills, where Buck had leased several square miles of pasture. Buck loved fresh eggs and had a pretty flock of Rhode Island pullets just ready to lay. As I loaded the ranch pickup and got ready to drive to California to receive the cattle, I crated the hens for him and placed them on my load. Somewhere along the way the door of the cage came open in the darkness, and when I got to Williams there was not a single hen left.

I left the outfit and my horses with a heavy heart, wondering if I would ever see them again. Some things I refused to leave behind were my big hat, boots, and spurs. Somewhere along the way I might get a chance to rodeo. I pressed them flat and managed to hide them through hundreds of shakedowns and inspections.

Not long after I finished Signal Corps basic training at Camp Crowder, Missouri, I got a weekend pass and signed up for the saddle bronc riding at a local rodeo. My hand shook as I signed my name as a contestant for the first time. I might have backed down, but I had done a little bragging around camp to my fellow soldiers, and several had come to the rodeo to watch me.

The hat was a little peaked, and the boots were mashed to where they put blisters on my feet, but I came out of the chute on a big white bronc named White Cloud and managed to hang in there for ten seconds to win some money, by virtue of the fact that there was not much competition.

At that first rodeo, I made friends with several local cowboys, who kept me informed of other rodeos and often picked me up at the camp gate and furnished transportation. The horses I drew in the bronc riding didn't buck half so hard as some I had been breaking in Oregon, and it didn't take me long to get pretty cocky about my talents. That ended quickly at Cassville, Missouri, where I drew a big Madison Square saddle bronc named Cheyenne from the famous Coburn string, who promptly sent me flying. For one thing I was scared and out of my league. I hit the big horse unevenly with my spurs and threw him into a spin I couldn't handle. It was like dropping a matchstick on a spinning phonograph record. I flew through the air and landed on what must have been the hardest, rockiest ground Missouri had to offer. That night I pulled two rocks as big as marbles out of my hip.

The next time I drew that big horse, I had already encountered the unknown and was determined to ride. He bucked straight across the arena, and I managed to make the whistle, but I took a beating. A week later, before my bruises were yet healed, I was shipped off to Camp Beale in California, where I was quick to locate a set of practice broncs owned by a local contestant, and managed to sneak out of the camp every evening to ride. There was no pickup man to help us off our broncs when the ride was over. Both of us became adept at stepping off the plunging animal and landing on our feet.

For want of better things to do, some of the soldiers in my outfit took to traveling with me, and when I won the saddle bronc riding at Marysville, California, in 1944, I had my first cheering section. My conceit didn't last long, however, for the next week I drew a famous Harry Rowell saddle bronc named Sontag. The rodeo was in San Carlos, California, in a small arena whose floor was covered with black peat. I paid my entrance fees, conscious that I had been reading about Sontag for some time as one of the great saddle broncs of the day.

I must have been overwhelmed by Sontag's celebrity, for I came out of the chute unable to psyche myself into thinking I could handle him. I was right. With no one to advise me, I took too short a rein. On the third wild jump, Sontag jerked me forward and I sailed out over his head and lit face-first into the arena dust.

I had bruises on top of bruises when I was sent overseas on the *Queen Elizabeth*, which had been converted into a troop ship. Landing in Scotland, we moved by rail to England, where we were hidden in an old cotton mill in Stockport, Cheshire.

While there, the army tried to toughen us up by making us bivouac in foggy, frozen parks, preparing us for the invasion of France. Some nights we were crowded back into the cotton mill, where we lay on cots, listening to the sounds of German buzz bombs overhead and dreading the moment the motors stopped and the bombs dropped on the villages below. By D day, June 6, 1944, we waited along the English Channel while planes roared above us and thousands of invasion craft, fraught with stormy seas, set off across the Channel for France.

Fate was playing roulette with our lives, and it wasn't much fun. Even as we waited, thousands of soldiers had already crossed the Channel and lay dead on the beaches. Soon I was huddled in the front of an invasion craft on its way to an army designation, Twenty Grand, near Le Havre. A little soldier named Nookie DeJovin called out, "Tonight's the night, men!" We all laughed, but the humor was only skin-deep. In no time we were back to eyeing the faces about us, wondering which of us would be lucky enough to survive.

There was a small galley on the boat, and as I passed it, looking for a latrine, I saw a whole case of canned Danish bacon. There was no one watching, and I stole every can I could pack under my coat. Moments later, as I watched the shoreline of France approach, an officer hollered, "Jump!"

Jump I did, thinking the water was only waist-deep. The heavy cans of bacon, the steel helmet, the rifle, and my ammunition took me down fifteen feet to the bottom. I struggled to remember in which direction the shore lay. Boots kicked at my helmet as other soldiers tried to swim. Suddenly a big hand grabbed my belt and pulled me along. It was my friend Bob Turner, a former heavyweight boxer from California. He dumped me choking and gagging in the sand. "I don't give a damn about you, Hyde," he said with a grin. "I was just trying to save your bacon."

I shared, of course. Turner stuck to me like a leech, and when finally we could stop and huddle behind a burned-out vehicle, we built a tiny fire and cooked strips of bacon. Never has food tasted so good since.

Assigned to Patton's Third Army, my outfit fought through France, Belgium, and Germany. We survived the Battle of the Bulge, the Ruhr Pocket, and the Rhineland campaign, helping set up signal communications for our fast-moving troops.

When the war with Germany ended, my outfit was sent to Arles, France, to await transport to the South Pacific, camped out in tents on a great plain south of Arles. I was homesick and bored. One day as I was passing the tent of the commanding general, I saw him sitting at his desk and strode in, saluting and introducing myself. I think I told him that I was a world champion cowboy, and I was volunteering my services to put on rodeos for troop entertainment in the nearby Roman coliseum at Arles.

All during my sojourn overseas, I had carried a photograph of myself winning the saddle bronc riding at Marysville, California. Now I plunked it down on the general's desk as the sole proof of my ability. For once in the army, things happened fast. The next day, there was my picture on that bucking horse on the cover of *Stars and Stripes,* the army newspaper, announcing that Dayton Hyde, the great bronc riding champion, would be producing and starring in rodeos for troop entertainment in the famed Roman coliseum in Arles. I read with fascination that I would be producing one rodeo and one Portuguese-style bullfight a week.

Fortunately for me, the town of Arles was not far from the Camargue, that wild, marshy plain that is the delta of the Rhone River. It was a land of wild horses and Spanish fighting bulls. During the German occupation, the animals had proliferated and were a perfect source of bucking stock for my rodeos.

Pursuing this lead, I was introduced to a Frenchman named Joseph Calais, who had been a leading bullfight impresario before the war. His wife had been a great bullfighter in the Portuguese tradition of fighting from horseback. Ordered to give a command performance for the Germans in occupied Paris, Mme Calais placed her darts in the bulls, each one breaking out in the red, white, and blue colors of free France.

The German officials watched in silent anger but did nothing. As M. and Mme Calais motored back to Arles, towing a trailer containing their famous bullfighting horse, Pronto, a lone German plane strafed them, killing Mme Calais and wounding Joseph and the horse.

Joe Calais was the perfect ally for me, but he had his price. He would furnish the wild stock for one weekly rodeo and bullfight if I would agree to produce and star in another weekly rodeo in the coliseum for his personal benefit. Not having any other access to stock, I agreed, and began canvassing the thousands of GIs in southern France for anyone who had rodeo experience.

The general himself got into the act by advertising that the army would pay one hundred U.S. dollars to anyone who could bring me a horse I couldn't ride. This produced a big gray coach horse who already had a reputation as a killer. During the occupation, the Germans had commandeered the horse, but he killed two soldiers before they led him the first mile, and the outlaw came back to the farm.

Beautiful as it was, the Roman coliseum at Arles was set up for bullfights, not rodeos. The worst problem was a low red fence called a *barrera* that circled the inside of the arena. On the plus side, it would enable the riders to escape the Spanish fighting bulls we were using for the bull riding. On the down side, the bucking horses might decide to jump the fence and would crash into the stone base of the stadium, pinning the riders between the fence and the ten-foot wall.

Lumber was scarce, but we finally scavenged enough planks to build a flimsy bucking chute. The horses and fighting bulls were brought in on trucks covered with a grid of steel bars so they couldn't jump out, and were housed beneath the coliseum. The day of the first rodeo came all too soon. That afternoon the coliseum filled clear to the ring of blue sky above it.

We had come up with a sprinkling of suicidal GIs who had rodeoed some before the war and were willing to try to ride those murderous fighting bulls. But I was apparently the only bronc rider, and it was evident that I would have to not only produce the show and keep it running smoothly but ride several bucking horses myself. I also took on the job of rodeo clown, whose responsibility it was to draw the fighting bulls away from fallen riders.

Joe Calais had loaned me an old Portuguese-style bull-fighting saddle with birdcage stirrups, and a cowboy hat more glamorous than my beat-up relic that had to be his prized possession. I swept out of the darkness beneath the stands riding Pronto, his big bay Portuguese Alter, circled the arena three times at a gallop, and slid the horse to a stop. The horse reared dramatically, pawed the air at the crowd, then dashed away again out of the arena.

I had little time to get my breath as I dragged the saddle over to the bucking chute and put it on the back of the big gray coach horse we had named Widowmaker. The French term for a bucking horse was *cheval sauvage,* and savage that horse was. By the time I got out on him, he had kicked apart half the chute, and he reared out bucking high and handsome. By the time he made it across the arena and was about to slam into the red fence, I stepped into the right stirrup and sailed into the air, landing on my feet. The man-hating horse charged back ready to eat me, but I managed to leap the barrera and save myself.

While we repaired the chute, I sent out a big Indian soldier named Chief Coser to do a trick rope act. The man was a little rusty and out of shape, but he bought us time. Soon we were firing out soldiers riding fighting bulls with loose-ropes around their girths, and the rodeo started getting wild.

The biggest wonder was that we didn't get anyone killed. Every time a bull came out and bucked off a rider, half the French youths in the stands would jump down from the parapets and start dodging the bulls, sometimes leading them right over the fallen rider. It would take fifteen minutes for the military police to clear the arena, and then the scene would start again.

After a few bulls, I would run in another saddle bronc and make a ride, then rush back to the chute to put another wannabe cowboy on a bull. Probably the most dramatic event of the day happened when I attempted to save a rider from being gored, and hit the bull across the face with Joe Calais's hat. To my horror, the sharp horn went right through the hat and left a big tear in the crown. I heard Joe Calais's scream of rage clear across the arena.

We survived the day, and by the next performance we had done some needed refinements to the chute and acquired some better cowboys from amongst the thousands of GIs who were camped in southern France awaiting transport to the South Pacific. If in that great Roman coliseum at Arles we performed countless acts of bravery, it was only because we hoped we would be injured and sent back to the States. Even death in the arena was preferable to being shot by the Japanese.

Eventually I shipped out of Marseilles on a troop ship bound for New Guinea, but the bomb dropped on Japan before we went through the Panama Canal, and we were rerouted to Norfolk, Virginia.

I was a long time getting back to the ranch. Stationed at Camp Polk, near Leesville, Louisiana, I paid the fine on a scrawny mare the sheriff had impounded and hid her in the empty barracks next to mine. Every night, I would slip out and lead the poor animal out to graze on orderly-room lawns, until finally she put some meat on her bones and began to prosper.

When at last I was discharged, I bought an army surplus truck, loaded my mare in the back end, and struck out for Oregon. I had spent all my money on the truck and needed gas money, so I ran a gambling ship, picking up hitchhiking GIs and running crap games in the back of the truck. By the time I reached Winnemucca, Nevada, I had just enough money to buy some hay for the horse and enough gasoline to head cross-country across the Black Rock Desert to Alturas, California, and Oregon. I was coming home at last to horses I hadn't seen in over two years.

Chapter Thirteen

I WORE MY UNIFORM WITH RIBBONS AND BATTLE STARS for a couple of days around the ranch just to remind my uncle that I had been away to war and he'd better not treat me as a kid anymore. He didn't seem to be impressed one whit. If I'd come home a general instead of a lowly private first class, it would not have mattered. All he needed was a hand who could take care of ranch emergencies and ride with him to open gates.

The first chance I got, I went to town and bought some western clothes. As to boots, I was too bound by tradition to tolerate store-boughts. I sent in the outline of my feet to Blucher Boots in Olathe, Kansas, and ordered some new boots, knowing that two years of marching in the army had altered the shape of my feet.

My feet weren't the only things around that had been altered by the war. So many of the old horses I'd come home to see had aged out during my absence. Sleepy, BK Heavy, Yellowstone, Roany, Spade, Badger, even the cranky colt I'd broken and named Brown Bomber after Joe Louis, every one of them was gone. Paddock told me they had all died of old age, but I suspected he had sent them down the road to be made into chicken feed. Had I not gone to war, I'd have fought for them, made damn sure they lived out their days on the ranch they loved. Whingding and Bright, the King Ranch horse Paddock had given me, were still around and were fixtures at Yamsi. Bright came up to the fence and nickered to me, but Whingding snorted, which was about all the real welcome I got. I hadn't been home ten minutes when Buck drove up in a new Chrysler and wanted me to open gates for him. The routine was the same, but I learned from the old man just how much the country itself had changed, although I had come home hoping things would be the same as I had left them.

For one thing, the Indians had sold out their reservation to the government for a national forest. It would not be long before the traditional grazing permits would be crowded out in favor of the recreationists. Down in California, folks were selling out their homes in an inflated market and paying such high prices for Oregon ranches that ranch kids had no chance of buying out their parents and were moving to town in record numbers.

And wages! When I'd left for the war, I was getting thirty dollars a month, thirty-five if I rode colts, and forty if I rode spoiled horses. That was more money than I knew how to spend. Now if I took a job with some absentee Californian, I could get four hundred or even more, but a good Visalia stock saddle would cost me over a thousand. What disturbed me more was that the out-of-staters were plowing up native grasslands, planting grains and grasses that wouldn't survive a summer frost.

Even the people were different. I walked down the main

street of Klamath Falls looking for familiar faces, but so many of those who had left for high-paying wartime jobs had never returned. And the ranchers I'd known since I was a boy? Gone now, and their houses either in ruin or occupied by strangers. For those who still toughed it out on the land and stayed home, maybe hoping that one day their kids would come back, their lives would never really be the same. Take our neighbors, the Brighthausens, for instance.

Longer than I can remember, there were Brighthausens living east of us on the edge of the desert, and most folks, before the war, grumbled at the thought the family might be there forever. The old man had raised a passel of little Brighthausens on other folks' beef, and I never rode that country for strays without feeling Brighthausen eyes on the middle of my back.

Back in the twenties, when other ranchers got pickup trucks, the Brighthausens stayed with horses. Folks said that the kids were raised with so much pine pitch on their britches they couldn't be bucked off, and for several years they furnished stock for the rodeos, big, rank horses raised on lava rock and strong grass, horses that mostly only Brighthausens could ride.

It took World War II to drive a hole in their ranks. There were Brighthausens that died heroes on the beaches at Normandy, and Brighthausens that died at Iwo Jima. Then old lady Brighthausen froze to death while out feeding cattle. After the war ended there were only two left: the old man, Lester, and his boy Jim.

The army had given Jim an education, and he came back to the Oregon country driving a good, if experienced, pickup truck, all ready to work the land as it had never been worked before.

It must have been a shock to old Lester to see that truck bouncing along over the old narrow wagon road from town, but the biggest shock of all came when Jim climbed out of that cab and walked toward the house with a beautiful, long-haired city girl by the hand.

"Pa, I want you to welcome my wife, Julie," Jim said, but the old man was so flabbergasted all he could do was stare. Then, without a word of greeting, he turned and walked away.

Lesser folks than Jim and Julie might have gotten into their truck at that point and driven away, but they stayed on, for Jim's love of the land was strong as was Julie's love for her husband.

Months went by, and Julie's cooking was so good the old man developed a little potbelly, and his pants sagged to the point where his Levi's wore out at the cuffs instead of the knees. He talked to Jim, but his disappointment in Julie kept him mean. As they sat at the dinner table, the old man would talk obliquely, telling about a neighbor who had married a city woman. "She never would close gates, thet woman. Let the bulls in with the yearling heifers one time, an' man, those heifers died like flies tryin' to birth those big calves."

"Pa, that's enough!" Jim would admonish, and the old man would shut up for a time, but gates were an obsession with him, and sooner or later he would tell how that city woman had left a gate open and let the steer calves they had weaned for shipment back out with the cows. "Don't know how thet man put up with thet woman. Musta cost 'im hunnerds of dollars in lost weight gatherin' those cows and separatin' those calves again!"

Before the war, Jim had captured a little bay mustang he named Buttons, and broke him gentle. Though older, Buttons was still around the place, and Julie soon learned to ride. Very often, when Jim needed help to bring in a bunch of cattle, it was Julie he took along with him, and left his pa behind. Jim loved the sight of his pretty wife galloping beside him, tiny-waisted and straight in the saddle, her long black hair streaming behind.

Despite his concern about gates being closed, old Lester never made things easy. His gates were simple wire gates with loops for latches, and when they were tight, it took a strong man or a wire stretcher to open and shut them. In order to keep Julie from riding the vast reaches of land she was beginning to enjoy, the old man took to tightening the gates so that even Jim grumbled about them.

Evenings, after Julie had ridden, the old man would drive out in Jim's pickup to check the gates as though determined to find one she had left open. Little by little, he tightened them so that soon Julie had no place she could ride alone.

October came; night frost settled in the hollows and turned the aspen leaves to gold coins. Julie rode long days with her husband gathering cattle off the ranges and came home exhausted, but somehow able to turn out a hot meal for the two men.

The gathering was only half done when Jim broke his leg riding a colt, and a pregnant Julie and the old man had to finish the job while Jim healed. There were times when old

Lester seemed almost ready to talk to her in more than monosyllables, but always his jaw would set hard and he would look off steely-eyed into the distance. Sometimes when Julie had to go through a gate alone, the old man would ride miles out of his way to see that the gate was properly closed.

Most of the cattle had been brought in off the ranges, and the first snows were settling on the mountains, when the old man devised a plan.

At the edge of the field which held the gathered cattle was a gate to end all gates. It had eight barbed wires, all so tight they sung like a harp in the autumn wind. The cold had caused the wire to contract even further.

It was all old Lester himself could do to open that gate, let them through, and close it again. One cold day they had passed through the gate and were riding the range together when Lester sent Julie to check on a hidden spring for cattle, and trotted back to the gate to add even more tension to the wire.

Finding nothing at the spring, Julie and Buttons trotted to the rendezvous point where she was to meet the old man. Old Lester lay flat on his back on the ground. As she approached, he raised his head weakly. "Miss Julie," he said, using her name for the first time, "I reckon I'm hurtin' real bad! Be a good girl an' trot on home fer help."

Wasting no time, Julie kicked Buttons into a gallop and pounded down the trail toward the ranch. Besides the old man's health, she had one worry, just how she was going to open and close that holding-field gate.

Once Julie had disappeared on her errand of mercy, old Lester rose to his feet, brushed the pine needles from his coat, and mounted his horse. As he rode on after her, he hummed a happy little tune. At last he was going to catch his daughter-in-law in the act. She might open that gate, but there was no way in hell she could get it closed! He would arrive just in time to turn back the cattle as they streamed out the gate to freedom.

But as the old man trotted out of the pine woods, he reined up his horse. Something was wrong! Her tracks led through the gate, but the gate was closed. Julie had outsmarted him. She had tied that gate up tight with her bra!

Chapter Fourteen

ROSE HAD WRITTEN ME ONE BRIEF LETTER when I was in France, and she seemed to struggle over what to say. She was in nursing school in San Francisco and missed her horse. There was a postscript I hardly noticed at the time, but it contained something of far more import than anything else in the letter. She said that quite a few folks from the reservation had been down to stay with her.

I might have known that Rose's family wouldn't give her up to a career without a fight. All that time overseas I had thought of Rose as a nurse, and used to search for her face in every busload of army nurses that went by.

I hadn't been home long when I passed one of her uncles on the streets of Chiloquin and asked for news of her, but he just stared at me and went on past as though he had never seen me before.

That night I had driven to a café on the highway for supper when I saw Rose herself. She came out of a bar and, for a moment, stood swaying in the full glare of my headlights. She held on to the hood of my pickup for a few seconds for support, then got into the backseat of an old car. The car roared off into the darkness before I had a chance to make myself known. Her face was puffy with bad health, and my heart near broke with the sadness of it all.

The next time I saw her, she was lying on the sidewalk in front of the general store. It was broad daylight, and folks were stepping around her as though she didn't exist. I picked her up in my arms, loaded her into my pickup, and took her to a motel, where I bought her a room and left her fully dressed and snoring on the bed. She was beyond recognizing me. I had believed in her dreams of becoming a nurse, and the collapse of those aspirations affected me with a depression as though her dreams had been mine. I found out later that her relatives had moved in with her in San Francisco and had partied so hard that Rose had been locked out of her apartment. It wasn't long before she gave up and came home.

The ranch seemed to be running pretty well without me, and my sojourn in the army had made me focus on getting an education. In June of 1947 I enrolled at the University of California at Berkeley, and was soon caught up in a world new to me.

The one thing that kept me from being homesick during my college days was the abundance of rodeos in California, but they were now different from those I had known before. Rodeos were getting to be big business with big purses. The competition was fierce and the cowboys hungry. No longer was the arena filled with big awkward ranch kids like myself. The new rodeo cowboys were often short in stature and super athletes. Many of them had never been on a ranch. To get into condition, they pumped iron and did roadwork. Instead of riding for pure joy, contestants entered several rodeos at once and rushed from one show to another by fast cars or light planes.

Time was, when a cowboy signed up for a rodeo he rode in parades up Main Street, danced with local girls at the fairgrounds, and mingled with the townspeople. Now cowboys rode their rough stock or roped, and a half hour later were on their way to another show. Big, long-legged cowboys like me could seldom compete successfully with the upstart competitors who had brought a professional slickness to the sport. I could pay my entrance fees and ride to the best of my ability, but what good was being in seventh place?

I was beginning to understand why my uncle objected to my interest in rodeos. For me, it would never be a paying proposition, and he knew that when I didn't. Rodeo was becoming big business, but with an increase in prize money came inflation of expenses. There was a jump in the cost of travel, medical expenses, and insurance. My uncle made a telling point with the question, "Who the hell is going to take care of you if you get crippled for life?"

Local ranch cowboys who came to town for a Fourth of July rodeo, hoping to take home some prize money, found it hard to compete with full-time professionals who rode a hundred broncs or bulls a season to their five or six.

I had ridden a few rodeo broncs in the army, but now, as a civilian, it was time to face reality and quit. I missed the thrill of setting down on a saddle bronc in a chute, hearing the crash of the angry bronc's hooves against stout planks, measuring off my bronc rein to match the horse, feeling the old familiar tightness of the saddle between my legs, the total concentration of getting ready, the soaring of that first jump out of the chutes when you felt the horse was never going to hit the ground again. I even missed the wake-up call of pain from old injuries as I rode, the jolts and twists, the trickery of a bronc trying to outwit you and put you down. I missed the roar of the crowd if you lucked out and managed to ride a tough one, or the silence, laughter even, or groan when the bronc won. I missed the thunder of hoofbeats and the feel of the pickup man's heavy muscles beneath his shirt as he galloped alongside, and the firmness of the ground as you swung down, safe at last, your ten-second ride finally over.* I missed the furtive glance back at the judges when you wondered if you missed keeping your spurs in the horse's shoulders that first jump, or if you had disqualified by touching the animal with your free hand.

I had saved a few dollars during the war, and to keep my mind off my failure as a bronc rider, I blew my savings on a Speed Graphic camera and some good lenses. I gravitated to being a rodeo photographer on weekends. This gave me access to the arenas, and created new friendships. I was bound and determined to take better rodeo photographs than anyone else, so I developed a technique of lying flat in the arena and shooting with a low camera angle. I took the risk of being run over by bucking horses and gored by bulls, but the low-angle shots gave greater elevation to the bucking pictures.

It also gave me the reputation for being a little bit crazy. At a Hayward, California, rodeo, I was taking photographs of

---

* The time needed to make a qualified ride was shortened to eight seconds in the mid-seventies.

a bullfighter clown named Slim Pickens as he was taking on Brahma bulls with a cape. I was immersed in my work, lying flat in the middle of the arena and getting great close-ups. Often the bulls would boil through Slim's cape and pass right over my body.

I was suddenly aware that Slim was standing there in the arena, looking down at me. "Kid," he said, "you're either the bravest man I ever saw or the dumbest."

The bull riding was about over for the day, but there was something Slim wanted me to do. "They're just about to turn out a fighting bull called Little Buck," he said. "I'm going to fight him Mexican-style with my cape, and I'd sure like to have a good photograph. You think you could get me one?"

Little Buck turned out to be a photographer's dream and charged Slim time and again. At times, the bull would threaten to gore me on the ground, but I lay still, and it would lose interest and go back to fighting Slim. Out of these shots, Slim selected one which would be reproduced in gold and put on a silver belt buckle. Slim was proud of that buckle and wore it for the rest of his days. For me, it was the start of a friendship that lasted nearly forty years.

The bulls continued to hold a fascination for me. I had learned cape techniques in southern France, and was intrigued by the beauty of well-executed passes. Every inch of the cloth seemed under control, and when I should have been studying I took out my Mexican *capo grande,* and practiced until my arms ached. When I went to a rodeo, I began to specialize in Brahma bullfight photographs, perhaps because the clowns always needed good publicity shots and there was no one else willing to stand out there and take close-ups. The clowns always assured me that they would take care of me, but time and again I would be all alone facing a charging Brahma. Often it took a cool head and a quick side step to prevent injury.

At the time, one of the great rodeo producers on the West Coast was a big Englishman named Harry Rowell, who had a ranch in the Bay Area at Dublin Canyon, near Hayward. Harry had come to America by jumping ship in the San Francisco Bay, and managed in time to put together a great rodeo business and a coterie of talented cowboys who made a sometime livelihood following Rowell's shows. Few promoters cared more for their cowboys than Harry, who paid the funeral expenses for many a destitute cowboy, and helped many another contestant through the off-season.

I got into clowning and bullfighting quite by accident, when a talented black clown named Felix Cooper was injured in an automobile accident and failed to show up for a Rowell rodeo. Harry had been watching me dodge bulls as a photographer and shoved me out into the arena to take Cooper's place.

When the clown is in the right place at the right time, he can save a fallen rider from serious injury by leading the charging bull away. That afternoon I worked close to the bulls and managed to save a few cowboys from being gored. It doesn't take much to turn a bull rider into a friend when you've saved his butt.

The barrel man that day was an old circus clown named Zeke Bowery, noted for his work entertaining crippled children. Zeke had several clowning contracts coming up in California and Nevada and needed a good bullfighter to team with him. He hired me, and soon on my weekends at the university I was not only taking pictures at rodeos but clowning. Even at the height of six foot five, I was fast on my feet and could succeed as a bullfighter where I had struggled as a bronc rider. Zeke and I worked in tandem to distract the bull. While I used a cape, Zeke had a heavily reinforced wooden barrel just big enough to hold his small frame that he would roll out into the arena during the bull riding and would duck into whenever a bull charged.

Harry Rowell owned a waspy Brahma bull named Twenty Nine, who had crippled the famous clown Homer Holcomb for life in Kezar Stadium. Twenty Nine had the sneaky habit of stopping just short of your cape and tiptoeing toward you. Since Brahma bullfighting depended upon the man using the forward motion of the bull, the man had little chance to escape. Twenty Nine could hit a man harder than any bull I ever saw. At a rodeo in Red Bluff, California, Zeke had just ducked into his barrel as Twenty Nine charged, and was clutching his handhold desperately when that big bull hit the barrel so hard, it hurled it through the air some sixty feet and over the arena fence, knocking Zeke unconscious. It wasn't long after that Zeke retired to gentler pursuits.

In 1948 I clowned with Slim Pickens in San Francisco's Cow Palace. I might have come out of that experience with a big head had not a *San Francisco Chronicle* reporter lauded my performance but misspelled my name.

In May of 1948, I took a photograph of a young man

bucking off Harry Rowell's Number Thirteen bull at the Point Reyes rodeo. The photograph became a *Life* magazine Picture of the Week, and the bull, a big, painted animal, was renamed Life Magazine. The bull was eventually sold to Christianson Brothers in Oregon, but he took the name with him.

Capitalizing on my opportunity, I rushed to New York and bluffed my way into the Time-Life Building to the office of the *Life* picture chief, Bob Girvan.

Mr. Girvan was out at lunch when I sat down before his desk. The desk was covered with photographs he was obviously considering. I acted fast when the man came in. Before he could ask me who I was, I blurted, "Mr. Girvan. You have such a flow of superb photographs going across your desk. How do you decide if a photograph is great enough for your magazine?"

Bob Girvan paused a moment in thought. "I have to have the feeling, 'My God, how did the photographer ever capture that!'"

"You're right," I said. "And doesn't that apply to great poetry, great literature, and great art?"

By the time the man had gotten around to asking me who I was, he knew he had been had. I pulled out a copy of my recent Picture of the Week and told him my ideas for stories I wanted to do for *Life*. He settled on a story on rodeos, gave me a check for $2,000 and all the photography supplies I could carry, and told me to meet him in San Francisco in September. I was off and running as a *Life* photographer.

That summer, I traveled widely with Slim, Margaret, and Margaret's little daughter, Daryle Ann. One week we would be in Billings, Montana, the next in Penticton, British Columbia, then Ellensburg, Washington. I blew the $2,000 advance on the story the first month and then was forced to live on what my slender rodeo and photographic talents could provide.

I did some suicidal things while photographing bulls, thinking that Slim Pickens was at my side and would save me. In Billings, Montana, the management cut the arena in half so the bulls would fight more. This gave me a far better opportunity for photography. I was doing some close shots of Slim fighting when a bull ran through his cape and headed straight at me. I turned and ran, leaping up the wire fence just as the bull nailed me in the seat of the pants. The wire panel sagged backward under my weight, pinning me on top of the animal.

Someone in the audience took a photograph, and there was Slim with a big grin on his face, pointing to my wallet pocket, showing the bull where to hit.

There was a camaraderie about that summer quite unlike anything I had ever known. I became friends with cowboys I'd only watched from afar. Often I traveled with Slim and his family. We camped together, cooked together, and shared our liniment bottles, for bruises were never far away. Between rodeos we spent restful days fishing western streams or on dude ranches such as the 4K in Dean, Montana, which was owned by Mickey Cochrane, the baseball player, and

Frank Book, who owned the Book Cadillac Hotel in Detroit. When we left the 4K, an underage girl stowed away in the backseat of my jeep, and I didn't find her until I had crossed two state lines and an international border. Scared, I drove like a madman to catch up with Slim and Margaret, and turned the young lady over to Margaret's keeping. The girl thirsted for an exciting, adventurous life like ours, but her father had arrived at the 4K just after we departed and was not pleased to find her gone. We heard over the radio that there was a manhunt on for the cowboys who had kidnapped her. We sent her home with a calf roper who was heading back to the States and failed to notice how young she was.

Before long, we were back in Montana. Slim and Margaret were invited to a barbecue by two all-time great lady bronc riders, Marge and Alice Greenough, and took me along. I had met their brother Turk, who was then married to Sally Rand, a famous exotic dancer. I had expected the Greenough girls to be big and tough like Turk, but here were two slender, lovely women who could have been pouring tea at a garden party.

The rodeos were hard work, but there was always time to relax. We would often arrive a day or two before a rodeo and give the horses time to get over the trip. Slim owned a big Appaloosa named Dear John and a pinto mule named Judy. John had been trained to buck on cue and could outperform most rodeo broncs. Riding him was part of Slim's clown act. Slim would rope and bulldog off the mule, Judy, to the delight of the crowds.

Often we would set up camp behind the chutes, amongst

the livestock trucks and pickups hauling horse trailers, greeting new arrivals as they came. There would be bronc riders braiding new bucking reins and working over their bronc saddles, bull riders working rosin into chaps and bull ropes, ropers swinging loops, and trick riders exercising horses and practicing runs. Inevitably there would be cowboys playing cards on a bale of hay covered with a blanket or playing pitch with pennies against the tire of a cattle truck.

Montie Montana, the great trick rider and roper, would generally be there caring for his horses, and I would wander over to help him. Montie's horses came first with him, and I don't ever remember seeing one of his horses that wasn't groomed to shine. Montie had worked in many an early movie with such greats as Tom Mix and William Boyd (Hopalong Cassidy), and if I hung around until after the work was done, I could usually tease out a story or two.

Montie and I remained friends for years. When he was in his eighties and still performing, he got two plastic knees. I was talking with him on the telephone and I said, "Montie, don't those artificial knees interfere with your lovemaking?"

"Only when I have to beg for it," Montie replied.

One morning, Slim and I had finished exercising the horses and were sitting in the shade of the grandstand, talking about fishing, when a cowboy named Bud Linderman came up. Bud was a brawny, curly-haired, hard-drinking, brawling black sheep of a great rodeo family, but he was a champion cowboy. According to Slim, Bud won most of his fights by jumping in quickly and putting out a cowboy's lights before the man even knew there was a fight happening. He had killed a detective at Madison Square Garden with one punch and was generally feared. Bud had been in a fight the day before, and the police had broken it up before Linderman had a chance to clobber his opponent. Now as Bud joined us, we could see his hapless cowboy opponent doing laps around the racetrack.

"What in hell's that dumb S.O.B. doing?" Bud asked Slim.

"He's training to whip you, Bud," Slim said with a grin.

Bud sat and watched as the cowboy did his laps around the track. Finally, when the man staggered by one more time, gasping for breath, Bud got to his feet. "I guess he's developed his muscles well enough by now," he said, and went down to the track to beat the daylights out of the cowboy.

In Red Bluff, California, I was mugging wild horses in the wild horse race. In this race some dozen or so wild horses were turned out all at once into the arena. Each horse wore a halter and trailed a lead rope. My two partners and I latched on to a horse that looked as though he could run a little and won the first day's prize money. On the second day, we scrambled to grab the same horse. We had a big Klamath Indian for an anchor man, who held on to the lead rope. My job was to mug, that is leap past the horse's front feet as it struck at me and grasp its head in a headlock, while the rider threw his saddle on the horse, cinched it tight, and vaulted up, ready to ride. The rider had just gotten the saddle cinched when Bud Linderman, who had been watching us, knocked him aside, leaped into the saddle himself, and rode the horse down the track to the finish line. We were all pretty disgusted when Bud won first money on our horse and our saddle.

I was working a rodeo in Lodi, California, when I next saw Bud. I was in a motel, and Bud was having a fight with his girlfriend in the room next door. Things were getting pretty loud. The only bathroom was down the hall, and as I went down the hall in my bathrobe, Bud grabbed me by the collar, dragged me into his room, and threw me on top of the woman in bed. I was scared and embarrassed, and the woman looked ready to die of fright.

What saved us both was that Bud started to build a cooking fire right on the rug, and suddenly alarms were going off, and the woman and I were able to escape in the smoke.

In Tucson, Arizona, Bud was raising hell in the drunk tank when the jailers subdued him by squirting him with a fire hose. Bud caught pneumonia, and soon died. Slim Pickens claimed that he went to see Bud in the hospital and could hear the man cussing three floors away.

"What the hell are you mad at, Bud?" Slim asked.

Bud grinned sheepishly. "I was just lying here reading the funny papers and caught myself siding with Dick Tracy."

Bud excluded, the Lindermans were a great Montana family. Bill Linderman was as fine a gentleman as I ever met, and one of the most talented cowboys that ever lived.

In September, I met with Bob Girvan and showed him the photographs I had taken. The editors of *Life* decided to do a Speaking of Pictures story of me taking pictures, and sent a *Life* photographer named Jon Brenneis to photograph me at the Merced, California, rodeo. The spread came out in the November 1, 1948, issue.

I had hoped that when the story broke it would make me a household name in America. A few days later, I began to wonder if *Life* had sent me the only copy of the magazine with my pictures in it.

"Kid," Slim Pickens teased not long after the event, "I reckon you just don't have what it takes to be famous."

In the spring of 1949, I was hired by a rodeo promoter who was trying out several carloads of Brahma bulls for rodeo potential. The very first bull jerked the cape out of my hands, and when I started to retrieve it off the ground where the bull was raking it with his horns, the animal charged from twenty feet away. I ended up with my right side in a heavy cast. For some months I had plenty of time to study, but had to learn how to take notes left-handed.

That November, Slim wanted me to work the Cow Palace with him again, but I was still wearing a cast. By then, Slim was well on his way to becoming a movie actor, and had lost track of the bulls and how they bucked and fought. I knew the animals well and agreed to be in the arena with him and spot the bulls.

I was a few feet from Slim in the arena when Twenty Nine came out of the chute. "Slim," I called. "This is Twenty Nine, the bull that hurt Homer. Let him go. It's not worth taking a chance."

That wasn't the thing to say to Slim. His big jaw tightened. "This one's for Homer," Slim said. As the rider bucked off, Slim stepped in to take Twenty Nine's charge.

The bull jerked the cape from Slim's hands and stopped, eyes snapping with anger. And then Twenty Nine started that deadly tiptoe. Slim tried in vain to get the bull to charge so he could sidestep the animal, but the beast kept creeping forward. Suddenly Slim was down under the bull's horns, and the bull was working a horn into his groin.

I forgot about my body cast and threw myself on the bull's stubby horns. Twenty Nine spun off Slim to meet me coming. I took a horn hard to my stomach that almost knocked the wind out of me but set me on my feet. Backing away, I slapped the bull in the face as I retreated. I ran just ahead of the bull's horns until he caught up and sailed me clear up into the stands. There was a look of surprise, then terror, on a woman's face as I dropped into her lap.

I was walking back to the dressing room after the performance with Slim and Margaret, when Slim put his arm on my shoulders and said, "Ya know, kid. I'm going to take you out tonight an' buy you the biggest steak that ever came off a steer."

"You sonova gun," Margaret muttered. "We had Slim insured for a quarter of a million dollars!" She was kidding me, I hope.

Chapter Fifteen

THERE WERE, OF COURSE, SOME REAL RANCH COWBOYS who did well in rodeos. Ross Dollarhide was one. He was raised on the P Ranch near Burns, Oregon. A natural athlete, he became a great saddle bronc rider and bulldogger, and a world champion. Ross and I were friends. We both came from Oregon, and when we would meet at some distant rodeo, we would exchange news from home. He died young, some say as a result of injuries from stunt work in Hollywood.

There were some fine ropers on the Klamath Reservation. Lawrence Hill, Friedman Kirk, Sandy Miller, and Dally Givons could win prize money at any rodeo in the country. There were saddle bronc riders like Buck Scott, Lee Hutchison, Jerry Choctoot, Dell Smith, Harold Hatcher, Phil Tupper, and Irvin Weiser, who could, on a good day, ride anybody's tough horse. But Monday morning would find them back at their ranches. With them, rodeo was a sport they loved to play but not a demanding career.

Sometimes I saw good friends die, like Kenny Madland, whose happy-go-lucky ways and infectious grin were snuffed out instantly by a Brahma bull. I had my own share of bruises and hungry times, but for me there were other considerations. When I rodeoed, it didn't matter a damn how many cattle and ranches my uncle owned. I was out there by myself, and if I succeeded or failed, it was all my own doing. Even as a rodeo photographer and bullfighting clown, I was living a life many a man would have envied, and I could count some great rodeo stars as friends, men like Jerry Ambler, Jack Sherman, trick roper Montie Montana, clown Homer Holcomb, Mel Lambert the announcer, Slim Pickens, and Ross Dollarhide. I hated to miss a rodeo for fear I'd be somewhere else when the excitement happened.

I got to meet interesting people with diverse talents, like Rex Allen, a western actor and singer who invited me to his home when he was throwing a twenty-fifth wedding anniversary party for Slim Pickens and his wife, Margaret. There was a story making the rounds that night about the time when Rex was waiting for a plane in the Los Angeles airport, and a fan rushed up and cornered him. "Mr. Autry," the man said, "would you please give me your autograph?"

Rex signed the autograph, "Gene Autry, who will never be half the cowboy Rex Allen is."

For a time, as a photographer, I knew all the great bucking horses — Badger Mountain, Snake, Five Minutes Till Midnight, Sceneshifter, Sontag, Miss Klamath, and Major Lou — and could tell you how most of the current crop of Brahma bulls on the circuit would handle with a cape. I wasn't good at many things, but there were cowboys walking around whose necks I'd saved from Brahma bulls.

Not a week passed when I didn't store up memories. Slim, Mel Lambert, rodeo clown Mac Berry, and I were at Red Bluff, California, and were visiting with Rex Allen, who was the featured entertainer at the famous rodeo. Some Chiloquin Indians I knew came up and invited us up to Oregon after the show to fish the Williamson River. "We got a place on the Williamson," one man said, "an' if you guys got a few days off, we'll take you fishin' on our property. Hell, we got trout on our place weigh fifteen pounds."

In those days the best fishing places on the river were Indian-owned, and unless a feller had an invitation from one of those Indian ranchers, he could maybe get gut-shot trying to slip into the river.

I had a sinking feeling, however, that the fishing trip might not turn out as well as we hoped. Once we had gathered any money coming to us from the rodeo committee, we headed north to Oregon, leaving Rex Allen to load his horses and music equipment and follow on his own.

If there was anything those rodeoing Indians knew better than team roping and bronc riding, it was how to throw a good party. As soon as they got back from Red Bluff, they got on the phone line and started calling relatives and friends.

"Hey," one of them said. "You better get your ass over here to our place. You know who's comin' here? Mr. Rex Allen himself. He's goin' to play his guitar and sing for us. You come an' don't forget to bring your old lady an' kids, an' say, bring plenty of whiskey an' beer."

By the time we arrived, the hay meadow along the ramshackle old ranch house along the river was already packed with cars. Kids a-horseback were galloping their mounts, playing games, and raising clouds of dust around the pickups and horse trailers.

I knew most of the folks, but Slim, Mel, and Mac, being celebrities, were introduced around, although they were impatient to slip away and fish.

"Where the hell is Rex Allen?" one of the hosts said, looking down the road. The man had worked for my uncle, and I knew him as someone not to be fooled with.

"He'll be here directly," Mel Lambert said, helping himself to a big venison steak off the barbecue grill. "He had to load up all his music stuff in his truck, but he promised to come right along."

Somebody hollered, "Chow!" and there was a general stampede of hungry kids to get to the head of the line. There were washtubs of beer and pop on ice, and some of the men had whiskey bottles stashed out in their pickups.

"That Rex Allen had better show up soon," one of the hosts growled angrily. "If he don't hurry, he'll have to go to town to a restaurant. Where the hell is he, anyway?"

"He'll be here," Slim promised, glancing nervously down the dirt road, which was already beginning to disappear in the gathering darkness. Someone popped a beer can near Slim's ear and dropped the can on the floor in a gushet of foam. Things were already starting to get out of control.

A woman started playing Indian drums and chants on a tape deck, and folks started dancing in a circle, while a few ladies corralled their children and left, as though they had seen wild parties like this start before. The Indian tape broke and was replaced by some forties swing. Mac Berry took a big lady by the arm and started jitterbugging, and pretty soon, she wound up kicking her legs up over her head as she danced. “I see you lookin’ at my panties,” she called out to Mel. “I made them myself out of flour sacks. See, they got Pillsbury’s Best on one cheek an’ Occidental on the other.”

“Hell, lady,” Mel muttered. “I thought I was lookin’ at the Grand Canyon of the Colorado!”

One Indian was not mixing with anyone. He was sitting over in the corner shadows, drinking from a pint flask and chasing it down with beer. He got up and lurched unsteadily over to where Mel was standing, watching the dancers.

“Damn that Rex Allen! When’s he comin’?” he snarled. “You know what I think? He ain’t comin’ at all! You white guys come here an’ eat up all our food, you drink our whiskey, you dance with our old ladies, an’ you tell us Rex Allen is going to come here an’ make nice music for the kids. You lie to us. Hell, you just want to catch fish in our river!”

The man moved closer to Mel, breathed heavily on his face, slipped a long hunting knife from a sheath on his belt, and held the point to Mel’s twitching Adam’s apple. “You lie to this Indian, I’m goin’ to take this knife an’ let yer hot blood run out on the cold linoleum.”

Slim Pickens rushed to rescue his friend. “He’ll be here,” Slim promised. “Rex gave me his word. He told me he’d be here, an’ he will!”

A big old barn cat with markings like a raccoon came in through a broken window and began chewing on a venison steak off someone’s plate. Mac Berry grabbed the cat from behind, and held it spitting and hissing in his arms.

“What you doin’ with my cat?” the man growled.

Mac had been having a good time and hadn’t the slightest

idea of the danger they were in. "I'm goin' to take him home and have him made into a coonskin hat," he joked.

"Damn you guys!" the man thundered, brandishing his knife. "You come here to catch our fish, you eat our food, you drink our whiskey, you dance with our old ladies, you want to kill our cats. An' you lie to us about Mr. Rex Allen comin' here to sing!"

"I think I hear him comin' now," Slim said, herding Mel and Mac toward the door. If they had intended to slip out the door and leave, though, the thought evaporated when a man motioned them back indoors with the muzzle of his rifle.

But there was no Rex Allen. The party got wilder and uglier. Women sat around the edge of the room with babies in their arms. Clouds of cigarette smoke hung in the air, then drifted out a hole where a falling tree had caved in part of the roof. A fistfight broke out between two men, and the mood went from ugly to perilous. The Indians who had invited us up fishing were nowhere to be seen.

"Well, I guess if we're goin' to get up early and go fishin'," Mel Lambert said, "we better go find a motel an' get some sleep. We'll try to find Rex in the morning an' bring him over for breakfast."

The man with the knife blocked the door with a chair and sagged down on it. "You guys ain't about to go nowhere until Rex Allen shows up," he said. "You think you can lie to us? Hell, he's probably halfway to Hollywood right now. You come here to catch our fish, you kill our cats, you drink our whiskey, you dance with our old ladies, you eat our food, you tell us lies. I'm goin' to take this knife and cut out the fronts of your pants."

"He's comin'!" someone shouted. "For chrissakes! Here comes Rex Allen!"

Indeed, out over the two-rut dirt road across the pastures, a vehicle was coming, shaking from side to side like a go-go dancer with the heaves.

Unaware of the havoc he had caused and weary from a long drive, Rex stuck his head in the door and located Mel and Slim. "Hey, you guys, I got lost. I'm glad I finally found this place. We gotta get up early to go fishing. Let's go find a motel and crash."

Someone grabbed Rex by his bola tie and set him down in a chair. Someone else raided his pickup for a guitar and thrust it into his hands. The big Indian stropped the edge of his hunting knife on his thigh. "Mister Rex Allen," he hissed. "We've been here all night waitin' for you to play music, an' damned if you ain't goin' to play!"

Rex glanced around the house, at the gaping windows, the holes in the ceiling, the falling plaster, and the door hanging from one hinge. The big man moved behind him with the knife. The old actor paused a moment as though he was thinking of his movies, where he had survived situations far worse than this. Taking up his guitar, he cleared his throat and started to sing "This Old House."

He got past "This old house ain't got no windows, this old house ain't got no doors," when the man with the knife shouted in his ear, "Hey, you goddammit. You come here to catch our fish, you drink our whiskey, you kill our cats, you eat our food, you dance with our old ladies. Now you make fun of our house! Maybe I should just cut your throat!"

Rex suddenly got it. He took up the guitar and began to play again. "I've got a little bitty baby in my arms. A little bitty baby, my, what charms."

All at once, the women got to their feet, took their babies in their arms, and began dancing around, singing along with Rex. The great Rex Allen had come all the way up from Hollywood to sing to them, and he was singing their song!

Chapter Sixteen

My uncle was having trouble finding men to help at Yamsi that year, and I gravitated back to the ranch from photographing rodeos. "Just temporarily, mind you," I told him. There were things with my life I wanted to do, and that didn't include building fence or opening gates for him. I had just gone to a ranch barbecue near Sprague River to look for someone to hire, when I saw a crowd gathered over near the barbecue pit. My old friend Al Shadley was lying on the ground, choking, with a piece of steak lodged in his windpipe. His dark face was purple, and no one seemed to know what to do.

Suddenly Rose darted out of the crowd, rolled the big man over, and caused him to expel the meat. Al's face went from purple to red, and pretty soon he was able to sit up and talk. Rose and I helped him over to a bench near the fire. The near-death experience had scared him pretty badly, and he kept promising first Rose and then me that he was never going to take another drink.*

"Do you think he can do it?" Rose asked when we were finally alone. "Just quit cold turkey like that?"

---

* True to his word, Al Shadley never took another drink.

"I don't know," I replied. "You'll have to ask someone who drinks. You, for instance. What do you think?"

She regarded me for a moment without answering. There was no sign of embarrassment. If she remembered anything of the day I had found her on the sidewalk, she gave no indication. "It's going to be harder for Al, since all his pals will be pestering him to drink with them. But when I think of the fright I saw in his eyes, I think he can."

"And how about you?" I asked. "Could you quit?"

She took me by the sleeve and led me away from the others, where no one could hear. "I already have," she said. She turned and regarded me. "We've never been serious about each other, but we've been like brother and sister for a long time. Can I trust you with my secret?" she asked.

I nodded.

"I'm going to disappear, and I don't want you to worry. The folks at nursing school have given me another chance, dependent upon my drying out. The only way my people will let me alone is if they think I'm dead. If I refuse to party with them, they give me the silent treatment and make me feel like hell. Pretending to die is rough medicine, I suppose, but I've racked my brain for another way that will work."

I had a little emergency money hidden in my wallet, my share of what a cowboy named Frank Mendes had paid me for mugging for him when he won the wild horse race at Salinas the year before. I offered it to her, but she shook her head.

"I'm fine," she said, her face flushing with embarrassment, and moved quickly back into the crowd.

It was starting to snow, and most of the Indian families were packing it up for home. I offered two or three Indians I knew a job, but none of them were interested. "Try old Coyote," one of them said. "He's getting out of jail tonight, an' maybe he'll want to work."

The jail was across the Williamson River from the town, and it was empty. It was snowing so heavily that the roof of the community hall was in danger of collapsing, and the prisoners had all been let out on work detail to shovel off the roof before the big Saturday-night dance. It was the last day of Coyote's sentence for being disorderly, and he was just putting in time. When I saw him, he was doing more leaning on his shovel than moving snow.

Once the snow was off the roof, they shoveled a path around the edge of the building to the big woodpile at the back, then went indoors to build a fire in the big double-barrel stove. Coyote remembered that he had a pint of whiskey hidden in the rafters from a previous dance, and by the time the crew was ready to quit and go back to jail, they were pretty happy. I made Coyote an offer, but he was in the mood to party and shook his head. I stuck around, hoping I might find someone else at the dance.

Coyote went back to jail with the other prisoners. He took a shower, grabbed a few belongings from his cell, plastered down his hair with Rose Hair Oil, drank the rest of the fragrant hair dressing, signed out with the town cop, and departed for the dance.

Hi Robbins had brought his rosewood baby grand piano down from Sprague River in the back of his pickup and got stuck in the snow before he could get it in the door, but there were plenty of sober folks about to manhandle the big piano into the hall.

By the time the dance had started, the barrel stove was cherry red, and the milling mix of dancers on the floor, about half loggers and the other half Indians, sweated in their flannel shirts and Levi's. Up in the bleachers, a host of ladies sat visiting or nursing their babies, bouncing with the beat of the piano and hoping someone would ask them to dance.

The door slammed frequently as men visited their pickups parked out in the snow, and now and then one could glimpse a flask sticking out of the rear pocket of a man's trousers.

I noticed one woman I'll call Cindy, who was the butt of jokes around the community. She never missed attending a dance, but there were rumors that no one had ever dared ask her out on the floor. It was even noised about town that once Cindy had gone for a swim in the Williamson River, and all the downstream fish had died clear to Klamath Lake.

That evening, having just come from jail, Coyote had run out of whiskey. He gathered some of his cronies and told them, "Hey, you guys. I gotta deal for you! You find me a pint of whiskey an' I'll dance a whole dance with Cindy over there."

"Aw, Coyote," someone laughed. "You wouldn't do that!"

"You get me a pint an' you'll see," Coyote replied.

His friends disappeared out to their pickups, and pretty soon they came back with a brown paper bag smoothed around a pint with no cork that had evidently been made up of the remnants from several bottles.

Coyote went behind a screen and gurgled down the whiskey, wiped his lips, tossed the bottle in a corner, and went off to ask the woman to dance. He pushed the woman round and round the room, being careful not to get her too close to the stove.

Once the dance was over, Coyote got together with his friends behind the piano.

"Hey, you guys," he said. "I got another proposition for you. You get me a fifth of Early Times, an' I'll take old Cindy out behind the hall an' make whoopee with her on the woodpile, an' I'll let you watch."

"Aw, Coyote," someone giggled. "You wouldn't dare!"

"You get me the whiskey," Coyote said. "An' remember, it better be Early Times, not that slop you brought last go-round."

It took a little longer this time, but pretty soon one of the men came back in the hall with the bottle, and Coyote was seen whispering in Cindy's ear.

He took her by the hand, led her out into the cold darkness, and as he went out the door, Coyote motioned to his friends to follow. Leading her around the building, he pushed her down on the woodpile, while Coyote's friends all gathered in the dark.

During the ordeal, Cindy happened to hear someone cough behind them, raised herself on one elbow, and said, "Oh, my, Coyote. We got audience!"

Once the game was over and the booze was gone,

Coyote got tired of being teased by his friends, and by the next week he got to thinking that fifth of whiskey was the most expensive he'd ever drunk. Cindy considered him her new boyfriend and followed him about whenever he came to town. Coyote even asked the sheriff if he would put him back in jail.

He was sitting in a local bar one night when Cindy came waddling in out of the cold. She was holding her hands together in front of her, with thumbs together, fingers overlapping, peeking between her thumbs and giggling as though she had something live in there.

"What the hell you snickerin' at, woman?" Coyote snarled.

But Cindy just backed up against a stool, peered into her hands, and laughed and laughed.

"What the hell you got in your hands, woman?" Coyote glowered.

"I ain't goin' to tell you. I got somethin' live in there an' I ain't goin' to tell. You gotta guess!"

Coyote turned his back on her and sipped his drink. But when he turned around again, she was still there, peering into her clasped hands and humming a tune.

"I tell you what, Coyote, you guess what I got in my hands, an' I spend all night with you in bed with no clothes on."

"Hell, woman," Coyote said. "I know what you got in there. You got an elephant."

"Oh, Coyote," the woman chortled. "That's close enough!"

The dances at the hall were mostly a wintertime affair, for in the summer men were busy working, hunting, or fishing. Hi Robbins kept his rosewood baby grand piano in an old deserted house he owned near Sprague River. I saw the piano often enough when I was loading truckloads of hay I'd bought from Hi. The roof was just beginning to leak on the finish. I begged Hi to sell it to me, but he said it was a gift he'd given his daughter, who had just passed away, and he wasn't interested in letting it go. That fall, some out-of-state deer hunters discovered the piano and loaded it in their pickup. It was a sad loss to the community, for that baby grand had brought music to many a dance, and was a part of local history.

Chapter Seventeen

My uncle made the most of my presence at the ranch by spending his days in Klamath Falls. He owned an office building which held several businesses plus a classy café called the Pelican Café. He was proud of that establishment, claiming that he wanted Klamath Falls people to have access to a San Francisco–class restaurant. Now in his late seventies, he served occasionally as a customer greeter, sitting at his own table up front where he could visit with his cronies as they dropped by. It was also a convenient place for the café management to display and sell the local newspaper. There was a dish on the table into which customers could drop the price of the paper. My uncle could often be seen obliging customers by taking money from the dish and making change.

I was on my way down Main Street to have dinner at the café with my uncle when an elderly Indian stopped me, shaking his brown fist in my face. "You ought to be ashamed of yourself!" he shouted. "You ought to be taking good care of your uncle instead of making him work at his age. I just saw that poor old man selling newspapers in the front of that café!"

My uncle didn't seem to appreciate the fact that I was putting in sixteen-hour days, seven days a week, trying to run a ranch that had formerly required five or six men. I was pretty desperate for help when an old character named Charley Tucker came to work for me. I might have been forewarned by the man at the employment office who phoned that he was sending Charley out. "Who knows? Maybe he'll turn out better than he looks," the man said.

"Better take his temperature before you send him," I said. "The last man you sent out here died before he got out of his car."

I was getting pretty tired of milking the old Jersey cow Buck liked so well and would have settled for anything just short of a corpse to help. Charley arrived late that afternoon in a beat-up old Chevy. Without so much as a by-your-leave, he unloaded his big green footlocker at the bunkhouse and moved in.

Without doubt he was the homeliest man I ever saw. As a scarecrow, he could have kept the crows out of fifty acres of corn. Six foot two, and so skinny his clothes flapped in the wind, he had a beat-up nose as crooked as a rail fence, large sun-scarred ears in a state of perpetual peel, faded blue eyes that appeared to have gone twice through Tuesday's wash, and the shifty, stubborn look of someone trying to claim an adventure he has just read in a magazine as his very own.

The one good sign, beyond a greasy cowboy hat, was his boots, which showed a wear spot where a stirrup had rubbed on a bunion. If they were indeed his boots, Charley had spent some time in the saddle.

I soon discovered that Charley's foremost trait was being proud. When I suggested that he take over milking the cow,

Charley pulled himself up straight and said, "Nope. I don't do that. Iffen I play with a set of tits, I want to hear giggles, not have someone kick over the damn bucket."

Charley and I soon came to detest each other, but we were kept together by our mutual need. He needed a job that kept him out of the bars, and I needed someone to help around the place. I guess that, beyond his hard appearance, the thing that really ticked me off was his stories.

He threw out his stories like challenges. "Call me a liar if you want, but when I was elephant man for the circus —" Another daily statement that never failed to clench my jaw in anger was, "You may not believe this, but when I was a cavalry sergeant in China, during the Boxer Rebellion —" Frequently he would ruin my day by the statement, "Back during the Second World War, when I ran the taxi fleet in Anchorage, Alaska —"

What really used to send me into the abyss was when he would talk about his sweetheart, Marge, "the most beautiful girl in the world." Marge was an obsession with Charley.

"One winter back during the Depression, my sweetheart, Marge, and I were hitchhiking across the Salt Lake Desert, heading west toward Wendover" dropped with a thud on conversations. I was convinced that Charley Tucker had never been loved by anyone, and that they even had to tie up a hind foot on his mother to let him suck. The woman named Marge had to be a fabrication out of a bout with booze.

Once, by sheer accident, I let him continue. "It was snowing hard and my little darling clung to me for warmth. I gave her my overcoat, determined to die for her if I had to.

Then, suddenly, out of the blizzard, a truck came by heading west. The trucker stopped, but he only had room for one, so I sent Marge with him and told her to wait for me in the next town. She blew me a kiss as she tossed my overcoat down. I came to a fork in the road just after that, and I knew there was no telling which way the truck had gone."

Charley lasted about a week that first time, before he got angry with me for questioning a story and left for town. About once a year he would show up again, and I would hire him because it was hard to find anyone who would work that far from civilization, and the man was cheap by virtue of the fact that sooner or later he would quit in a huff and walk the thirty miles to town without bothering to collect his pay.

Sometimes the employment office in Klamath would send him out in response to my plea for even a slightly warm body to help feed the cattle; sometimes he would come dragging in on his own afoot with the story that his danged pickup had just quit him down the road. During the fifties he worked for me and quit some forty times, but always he would show up again, looking older and more dissolute with every arrival, and he would hardly set his feet under the dining room table than he would begin, "One time when I was elephant man for the circus —"

It got so he left his footlocker in the bunkhouse even as he quit in a towering rage, and one steel cot became permanently known as Charley's bunk, its tattered mattress sagging in the middle with the ghost imprint of a body too long for the bed.

I saw more of Charley on Mondays than any other day, for it was always the lot of Charley's employers to get him out of jail after a Saturday-night brawl. It wasn't that Charley drank so much. As I see it, he never had a chance to more than sidle up to a bar and shove an opening sentence right into the middle of someone's conversation. Like, "I don't mean to interrupt, gentlemen, but one time when I was driving a twenty-mule team in the cavalry —" The fight was on just as soon as someone inferred that Charley Tucker was a damn liar. No one, including me, reckoned that a man who looked like Charley could have done much of anything.

By virtue of Charley's brawls, I came to spend Mondays in town, banking, buying new parts for worn-out farm machines, and stocking up on groceries. I paid Charley's fine last, leaving him to cool his heels in the drunk tank until most of the day was gone. Thus there were two glorious days of the week, Sunday and Monday, when I didn't have to put up with him.

It wouldn't be long before Charley would be as sick of me as I was of him, and our divorces from one another always came easy. I had only to sit down at the breakfast table and say with a sigh, "Sure is a lot of work to do around this ranch." I would then proceed with a list, "Got to fix the K Davis corral and check the fence around Calimus Field, get the baler and swather ready for haying, then move about a hundred fifty of those heifers down to the Barfield."

The thought of all that work would put Charley in a depression, and minutes later, he would be on his way to town.

Not that he didn't work hard when he was here; he just couldn't stand the pressure of thinking about it.

Once we were building a new pole gate at the K Davis corrals, Charley and I, and the ladder I provided was admittedly pretty rickety. I was in a hurry to get the gates built so we could work cattle, but Charley insisted on building a ladder that would hold. He spent one precious day peeling poles and building a good, strong product, and had just leaned the ladder up against the towering gatepost and stepped back to admire his handiwork when a freak wind came down the Bull Pasture Draw and blew the ladder back off the post.

I suppose I was already angry that the gate hadn't been completed and the corral wasn't ready. When the falling ladder glanced off my shoulder, I whirled in pain and anger and gave it a kick. I hardly touched the damn thing, but it broke in two, and there was Charley climbing the west fence, heading through the timber to town. It was months before I saw him again.

Nothing ever went well for Charley. The next time he came back, I sent him down to rebuild a fence at the north end of the ranch, and he mistakenly got on my neighbor's property and permanently planted a couple of thousand dollars' worth of my steel posts.

Once, a movie company arrived, needing some footage of cattle being driven through beautiful ponderosa pines. Charley fed the director a good line about how he had ridden for this big outfit and that, and proceeded to borrow a good horse I had, named Matador, for the movie scenes.

"When I was a sergeant in the cavalry —," Charley said

as he mounted the big black horse, but the director was busy working with his crew.

I suppose I should have mentioned to Charley that Matador was a one-man horse and could buck better than most men could ride, but Charley kept at the director's elbow, telling him one story after another, and I had no chance to tell him about Matador without hurting his feelings.

Contrary to expectations, Matador let Charley ride, and soon the man and his horse had gathered a small bunch of cows and calves and were driving them toward a beautiful crystalline spring that gushed from a hillside and wandered down through a grove of ponderosa pines.

The photographers had set up cameras on the hillside, and the director shouted directions to Charley to drive the cattle on past. The cattle were doing fine, drifting down an old trail, and Charley had little to do but follow them at a walk, but he began to dash back and forth, waving his arms dramatically and hollering at the cattle to "git" down the trail.

Standing just out of camera range, I quietly suggested to Charley that unless the cameraman had some really fast film, it might be better to let the cattle drift quietly at a walk. Charley glared at me, kicked Matador in the ribs, and went flying at the cattle with a vengeance. That was enough for the old horse. He ducked his head and drove Charley's face into the dirt right in front of the camera.

"Cut!" the director called as the old man rose sputtering from the dirt. "Get back on, and we'll try that bucking scene again!"

We never did the scene again, of course, for Charley

caught his horse, tied it to the fence, and took off walking for town.

When I next saw Charley, it was late in the fall, when the harvest hiring was over and jobs were scarce. I heard the rumble of an old automobile out on the main road and moments later saw headlights lurching through the trees as a car came rumbling down the dirt road into the ranch. When the driver parked in Charley's old spot under a pine tree next to the bunkhouse, it was obvious the old boy had come home.

That evening, Charley was back at his old place at the table, and his windy stories were once more of the background murmur as we ate our supper.

"Back before the Second World War," Charley began, "when I was up in the Yukon prospecting for gold —"

I dropped a serving dish in the kitchen, and the story was never finished.

About a week later, a cattle buyer from California drove down the road, hoping to make a deal on the fall calves. He was driving a big, fancy Chrysler, and his wife was a beautiful, well-dressed lady who wore high heels and was obviously out of place in such rustic surroundings. The man was attentive to her needs, and it was obvious that they adored each other.

I invited them in for lunch and soon had a table of food ready for them and my crew of one. I rang the chow bell for Charley, and we were seated at the table when he came over from the bunkhouse.

I was about to introduce Charley to the guests when he stared at the woman, spun around, and departed. We had our lunch, but Charley never reappeared.

After lunch, when the buyer was making phone calls, I walked over to the bunkhouse and found the old man sitting on a block of wood.

"Charley, I reckon you knew that woman," I said.

"Yep," Charley said slowly. "Back in Anchorage, Alaska, during the war, I was running a fleet of taxis, and she was running a fleet of whores. Looks like she's got herself a fancy husband now, and I didn't want to recognize her and have her old man ask where we met."

"Charley! Charley, you old devil, how are you?" the woman said, coming up from behind. "You don't need to worry about my old man. Hell, it was my money set him up in the cattle business. We've been married for years. Ain't much about me he don't know."

Charley stayed on that fall pretty well, and even agreed to stay on during the winter as caretaker when the cattle were shipped to northern California pastures for the winter. He would be snowed in for months, but I didn't worry about the man. There would be no one around to take offense at his stories, no one around with whom to do battle.

In the kitchen was the old crank telephone, connected to the world by a single line from tree to tree. To get out, one first had to ring an operator in town. Since Charley would be snowed in for long periods, I made arrangements with a woman in town to phone Charley every day to make sure he was safe.

Charley and the woman had three things in common. They both loved to talk; they were both lonely people; they were both homely as sin. Before I knew it, they had fallen in love sight unseen and were thinking of getting hitched come snowmelt in the spring.

There was a time that winter when snow came drifting down off Calimus Butte and the road was totally obscured. Even the chickadees and gray jays forsook the ridges until spring.

I was in California with the cattle and seldom got a chance to phone, but sometimes I talked to the woman, and sometimes, even, I managed to hear Charley's voice across the miles of snow. Charley's stories took her to a new world as he traveled with elephants in the circus, panned for gold in Alaska, and drove twenty-mule teams in China.

I first knew the snow had melted on the road when the sheriff called me in California to let me know that Charley was drunk and in jail.

I realized that there was no one left at the ranch to care for the livestock and drove two hundred miles north immediately. "What happened, Charley?" I asked as I paid his bail. "I thought you and that woman were going to get hitched."

Charley looked to be in bad shape, as battered and bleary-eyed as anyone I ever saw.

"It was that danged woman's fault," he said. "Over the phone she said she weighed about ninety-three pounds. She got to sounding more and more like Marge, and I got to picturing her as movie-star pretty. I won't say how big she was, but when she got to the ranch in her old pickup truck, the springs were all broke on the driver's side."

Either Charley's disappointment was too great or he had been in one place too long, for he disappeared from my life.

I got a few cards from him over in the Winnemucca, Nevada, country where he was riding for a big cattle outfit, then from the Sacramento Valley in California where he was working the rice harvest.

For a couple of years there was no word at all. Then one day I received a letter from a rancher in Colorado saying that a man named Charles Tucker had died on his ranch and had been buried there on the land. He had found my address in Charley's wallet. Charley had some back wages coming, and the man wanted to know where he could locate Charley's next of kin.

I didn't know, of course, but I went over to the bunkhouse and pried the big brass padlock off Charley's foot-locker, hoping to find some clues about his family. There on top of some clean but faded military uniforms were photographs of a handsome sergeant in uniform driving a team of twenty mules in China, Charley putting a group of circus elephants through their paces before a cheering crowd, Charley sluicing gold in Alaska, and the hauntingly beautiful face of a woman looking out of a photograph signed, "To Charley, with all my love forever. Marge."

Book Two

Chapter Eighteen

NOW AND THEN, WHEN RETURNING TO YAMSI from photographing a rodeo, I would ride south over the tablelands and run into Bart Shelley, driving his tobacco-stained pickup, and ask him about my favorite bucking horse, Blackhawk. It would get the old man to talking, since the horse was the center of Bart's world. Blackhawk ran free with Bart's horses on the tableland, and it was only once or twice a year that the old man would run his horses in for haying or for weaning colts. Then, according to his mood, he might put a halter on the old bronc and lead him over to Beatty for a Sunday rodeo to give some unlucky cowboy a chance to ride him.

The horse was a lot better than the local competition, and in the dozen or so times I saw him buck, I only remembered seeing one cowboy make a qualified ride. That was Ed Donovan, whose ride I had photographed as a boy. Blackhawk leaped so high you could have driven a pickup under him, and turned flat on his side in midair in a sunfish. I would have sworn the big animal was going to fall on Ed's leg.

The next jump was mightier than the first, and Blackhawk squealed in anger and kicked toward the heavens. The horse's back hooves sent a rain of gravel rattling down on the cars parked outside the arena fence. The crowd lurched to its feet in excitement, and the cheers began before the ride was half over. When Ed Donovan made it to the whistle, I felt as though my old friend Blackhawk had been cheated somehow, and I stayed back in the catch pens with the old horse until the rodeo was over and Bart trotted up a-horseback to lead the animal home.

The rodeo promoters up and down the coast all wanted that horse desperately, but Bart wouldn't sell. Potential buyers would come from as far away as Montana and stand, looking longingly at the long-bodied athlete as he stood in Bart's corral, so engrossed that they failed to notice that Bart was staining their fancy boots with tobacco juice. The old man could spit through a log fence and hit a man's boots every time.

"By gawd," Bart would say, "thet Blackhawk is shore some buckin' horse" (spit).

Every year, Mac Barbour would try to buy the horse for his bucking string, but with no more luck than the others. I hated it that the animal was getting on in years but had not yet made it to the big time. The big black was twenty-two years old when Bart finally died and Barbour was able to get his hands on the animal.

For some reason the horse had always had a liking for me. Whenever I would see him standing in Bart's field, I would get out of my pickup for a visit. The horse would leave his group and come over to the fence, and I would stand rubbing the arch of his neck as he nosed my pockets for grain. At Yamsi I would lie awake nights dreaming that I was at some famous rodeo or another and had drawn Blackhawk in the finals of the bronc riding. He would be nice to me, of course, and let me steal a ride. I would step off into the air just before the whistle to let Blackhawk's record of buck-offs remain unblemished.

The Pendleton Round-up in Oregon was, and still is, one of the greatest rodeos in the United States and Canada. Only the best saddle broncs in the country are selected for the finals of the saddle bronc riding event at the end of the rodeo. Blackhawk was one of those selected. At twenty-two years of age, he had finally made it into the big time. Pendlcton is some four hundred miles from Yamsi; I would have walked the whole distance rather than miss the finals.

Jack Sherman was one of the saddle bronc finalists that year. He had drawn another of Mac Barbour's great bucking horses, and Mac, in a fancy white western shirt and red bola tie, was bustling up and down behind the chutes, peering through the slats at the horses, making sure the finalists were there with their riggings ready to ride. Jack's animal was all saddled and standing quiet in the chute, and Jack himself, with a new black hat and his usual crooked grin on his face, was psyching himself up to win.

I was helping behind the chutes when it came time for Jack to ride. Directly above the bucking chutes was an extension of the grandstand, heavy fir planks that were now sagging with the weight of three very large ladies who had drunk more than their share of beer. Their bare thighs overflowed the planks and cause a giggle or two from us cowboys who were working beneath them.

"Look at that, will you," Jack said with a grin. "I'd better come out on this old horse fast before those planks break."

Jack had just started to climb down on his mount, and was measuring off the proper length of bucking rein, when Mac Barbour reached up with a couple of electric cattle prods and touched the ladies on their bare flesh.

With a horrendous scream, the ladies grabbed each other and lost control of their bladders. There was no stopping the flow. For the next minute, buckets of urine poured down over the chutes. Jack Sherman, his big black hat and rodeo shirt sopping wet, hit the ground in a towering rage. Mac Barbour took off down an alley as fast as his stubby little legs would carry him.

"Now Jack," he pleaded. "Now Jack, be careful of me! I've got a real bad heart!"

When Jack came out of the chute a few moments later with a borrowed hat and shirt, he was still mad enough to make one hell of a ride.

Blackhawk had been drawn in the finals by a Washington State cowboy named Alex Dick, who was known to be a hard man for a horse to buck off. To Alex, the big black horse in the chute must have seemed like just another horse. He sat watching from his perch on the chute as Ross Dollarhide came out on a little bronc named War Paint. That feisty little paint horse reared and came over backward on Ross, but the big cowboy rolled away and escaped injury.

It was now Alex Dick's turn. The announcer described Blackhawk as one of the greatest bucking horses ever to come out of Oregon. I looked over the top of the chute to see the horse's black eyes snapping with anger, then glanced at the brown features of Alex Dick, and saw the intensity of his concentration.

I was standing in the arena when the chute gate opened. On the first jump Blackhawk leaped so high I could see the face of one of the judges beneath his hooves. Alex Dick's neck snapped back as the animal hit the ground. On that next jump Blackhawk turned sideways in the air in the sunfish he was famous for, then, as his black hooves pounded the arena floor, he ducked back underneath himself and changed directions with a lurch that would have sent most cowboys flying. But still the Indian cowboy spurred from the shoulders of the horse to the cantle board of his saddle.

I scampered past one of the judges and kept pace with Blackhawk, screaming encouragement. If Alex Dick managed to hear me, he probably thought I was cheering him on. In reality, I was encouraging the horse.

Suddenly there was the whistle, and a pounding of hooves as the pickup man galloped past me and raced alongside Blackhawk. Alex Dick grabbed the man's shoulders and swung to the ground. The crowd was on its feet, cheering, and I stood stock-still in the middle of the arena as Alex walked back to the chutes in triumph.

That evening, I sat with Blackhawk and the other saddle broncs behind the chutes. The big horse stood calmly munching some grass hay as though the afternoon had never happened. I cupped one hand over one of Blackhawk's sweaty eyelids and scratched him, while the animal raised his nose in ecstasy. I could not know that I would never see him alive again.

On the way back to Klamath Falls from Pendleton, Mac Barbour's big livestock truck, driven by my friend Jackie Houston, tipped over, and Blackhawk had to be destroyed.

I was back at Yamsi when I learned about Blackhawk's accident. I loved that animal and always dreamed that someday I'd make a ride on him. Now it would never happen. Something went out of me with his death, as though a family doctor had told me I'd never ride a horse again. Bart Shelley never did the horse a favor by owning him so long. At twenty-two, Blackhawk was past his prime when he got his chance to prove himself at a major rodeo, but some of us will always tout him as one of the great bucking horses of his day.

In the late fifties, my uncle died, and I felt depressed. It was the end of a great era at Yamsi, and the place would never be the same without him. Often, when I came into the Yamsi kitchen, I would expect to see him sitting at the breakfast table reading the *Saturday Evening Post,* but he wasn't there anymore.

The morning I picked up his ashes at a funeral home in Klamath Falls, it was twenty-five degrees below zero. I was driving home to Yamsi and was just passing the Chiloquin airport when I saw a rancher friend, Bert Stanley, working at thawing his little single-engine airplane, trying to get it to run. I had been missing some cows on the range west of the ranch, and stopped to see if Bert would fly over the frozen countryside with me and help me locate the cattle. It would also be a good time to give Buck's ashes a good scatter over the land he loved.

Bert had finally gotten the engine to turn over despite the cold, and was letting it warm up when I asked him to fly me over the range. Bert was missing a few cows on his Hog Creek ranch and agreed to take me flying.

"Do you mind if my uncle goes along?" I asked.

"Not at all. Where is he?"

"He's in this box under my arm."

The idea took several minutes and a pint of coffee to get used to, but eventually Bert agreed.

We located three of Bert's cows on Wocus Bay, then flew over Skellock Draw, crossing the long snowbound ridges to the valley of the upper Williamson River. There in the upper reaches of Haystack Draw, we saw some of my cattle grazing on bitterbrush in the pines.

As we gained elevation to fly back to Chiloquin, I shouted to Bert above the roar of the small engine. "Would you mind?" I asked. "Would you mind flying over the ranch so I can give Uncle Buck's ashes a scatter?"

Bert's tanned features turned a little white at the suggestion. I could see he was bothered, but I kept on. "He was a historical character in these parts," I said. "You and I both owe him something."

The little plane climbed higher and higher over Yamsay Mountain, until the ranch was only a long white ribbon of snow amongst the pines. We were now higher than it was safe to fly, and we were both getting a little giddy from the cold and lack of oxygen.

Soon Bert began to grin. "I used to be a pilot in the war," he said. "Tell you what. I'll put her into a dive over the ranch, and when I holler bombs away, you get your size fourteens in the door and give the old gentleman a scatter."

Suddenly we were in a screaming dive toward the ranch, which was swiftly getting bigger and bigger. "Bombs away!" Bert shouted above the din.

I had already pried open the top of the wooden box and opened the thin cardboard container within. It took all my strength to open the door against the rush of wind, but I managed to get my cowboot wedged in the door and emptied the box into the slipstream.

"We're running into a cloud bank!" Bert shouted.

"Cloud bank, hell," I shouted back. "That's Uncle Buck!"

Half of the ashes had been swept back into the plane, and now we were blinded and choking to death as Bert tried to see out the window.

Just in time, he pulled the plane out of its dive. *Snick, snick, snick* went the tops of pine trees as the landing gear cut them off.

Bert was as pale as though he had grown up under a rock as we roared up over the ranch. Far below, I could see my hired hand, Al Shadley, driving the team and wagon with a load of hay out over frozen meadows dotted with cattle. I still had the wooden box and half the ashes left. "One more time," I pleaded, sweeping up all the ashes I could reach in the cockpit.

We made one more run, and this time I threw out the whole wooden box, ashes and all. I tried to watch as the yellow cube hurtled down toward the meadows, but lost sight of it behind the plane.

Two hours later, I was back at the ranch cooking venison over a roaring fire in the stove when Al came in for lunch. For a long time he was silent, brooding into the steam arising from his coffee, and then he said, "You know? The damnedest thing happened to that Jersey cow your uncle thought so much of. I was feedin' the cattle an' found her layin' dead out on the feed ground. Wasn't a thing I could see wrong with her except a pile of boards fell out of the sky and hit her right smack between her horns."

After Buck's death we still cooked in the kitchen of the big house, but when cold weather came, I grew tired of keeping fires burning in the kitchen just for me. The bunkhouse had an ancient box stove that took a pretty good log, and I soon moved in both to stay warm and share company with Al, who never seemed to run out of stories.

That stove had a personality of its own. It was forever either getting hotter or cooling down, and the monotonous tick it made as the temperature fluctuated paced the flow of cowboy conversation like a metronome. Sometimes pockets of gas in the pine logs would explode and startle us into thoughtful silence; sometimes amidst the crackle and pop of a healthy fire, moisture would ooze from the logs and spit and hiss as it boiled away in the heat. Sometimes the stove rumbled and shook as the logs collapsed into coals. The old cast-iron stove was held together with a couple of bad welds where some forgotten cowboy had attempted to mend the cracks. Firelight danced upon the rough boards of the floor, and a sudden downdraft in the night could cover the cots and tables with gray ash. Cigarette smoke hung in ghostly layers in the superheated air, and the talk in these postwar years was often desultory. Except for Al and myself, there was seldom anyone there with a sense of local history, or with much experience with cattle and horses.

Often Al and I talked of the old days when cowboys looking for jobs traveled light, arriving at a ranch with little more than their saddle, bedroll, and, rolled in the blankets of the bedroll, a change or two of clothes. If they wanted a chair in which to sit before an evening fire, they picked up a couple of old boards around the ranch and, without benefit of nails, whittled a crude chair which could either be left behind or knocked down, wrapped in a bedroll, and taken to a new job.

Until World War II, these cowboy chairs were a common item in line camps and bunkhouses throughout the West, along with orange crates and apple boxes which were handy for storing rain gear, spare socks, western novels, and other necessaries. After the war, these chairs were the first traditions to be lost and, most likely, made a ready source of kindling for winter stoves.

Lost along with the crude chairs was the art of storytelling, once kept alive by old cowboys who could spout a verbal history of each ranch around the bunkhouse stoves. Often, as I sat around the bunkhouse waiting for bedtime, I would glance about at empty cowboy chairs and try to remember old friends who had sat there spinning magic tales.

One favorite of mine was a cowboy named Dick Blue. Dick's ancestors were eastern Indians who had migrated west and settled in Washington State at the turn of the century. They homesteaded a small ranch along the Columbia River, not far from the Canadian border.

Dick would add a log to the bunkhouse stove and talk about his relatives as the fire began to roar. "I was never sure who my real grandma was," Dick told me. "The old man always had plenty of women around to do the work and never seemed to show much affection for any of them. I remember one day about noon, one of the women came in sniffling, holding on to her arm.

" 'What's wrong with you, woman?' the old man snapped.

" 'I was out chopping wood and got bit by a rattlesnake,' the woman whimpered.

" 'When do we eat?' the old man demanded."

After lunch, the old man harnessed a team and drove the woman out to the highway, where she could hitch a ride into town. It was months before Dick saw the woman again.

Dick finally settled on one of the women as his grandmother, mainly because she was nicer to him than the rest. To show affection for the woman, Dick and his brother saved up and bought her a gasoline-powered washing machine. To start the engine, there was a big foot crank. She still had to pack water from the river, but the job of doing the old man's wash was now lots easier. One stroke of the crank and the engine would shoot blue doughnuts of exhaust and start putt-putting away, and the drum containing the clothes would turn round and round until the wash was done. The boys, pleased that they had done something nice, headed home.

Just about that time, the old man traded some horses for

a battered 1930 Model A Ford pickup. It wasn't running too well, so he borrowed the spark plug out of his wife's new washing machine. The pickup still didn't run very well, but the woman discovered that the washing machine didn't run at all. But she noticed that every time she mashed down on the foot pedal, the barrel of the washer would turn around one quarter turn. From then on, that is how she powered the washing machine, pushing down on the cranking pedal with her foot until the washing was done.

Dick's grandfather was a good enough worker, but he had a real resistance toward planning a job ahead of time. Whenever Dick would suggest that the old man plan his projects better, he would reply, "When you build a roun' corral, it don't matter a damn which post you set first."

Dick's partner, Claude, was taking a pack trip into the Washington Cascades. He and his brother were packing with horses miles away from civilization when the brother began complaining about a pain in his heart. They were just breaking camp when the brother said "Claude! I think I'm having a heart attack!"

Claude said, "You see that pine log over there? Go drape your body over it with your head hanging down."

"Will that help my heart attack?" the brother asked.

"Well, no," Claude admitted, "but if you die, it will sure as hell make it easier to pack you out of here."

Chapter Nineteen

THERE IS SOMETHING ABOUT RODEOS — the dust, the roar of the gaily shirted crowd, the camaraderie between contestants who travel together and help each other, the near-death experiences, the all-night drives from one town to another, the great horses, the hoofbeats of charging bulls and the clanking of brass cowbells hanging from bull ropes as the bulls buck, the friendships with bull riders whose life you've saved while risking your own. The hospitality of folks in rodeo towns and the adoration of fans. But in the life of every rodeo athlete there comes a time to retire.

Slim Pickens was getting well known as an actor. He was the cowboy who rode the bomb in *Dr. Strangelove,* and later would play a sheriff in Mel Brooks's *Blazing Saddles*. With every successful part, his studio was more and more reluctant to let him risk his neck clowning. He held the clowning contracts at some major rodeos and offered them to me, since he was now unable to honor them. It was a chance of a lifetime, but not something I could abandon the ranch for. I hadn't practiced with my cape in months and knew I had to come to grips with reality. Sooner or later I would face a smart bull I couldn't handle, and I'd end up injured or even dead. Also, if I went back to clowning, I would have to get funnier in order to endear myself to a crowd. With his big, round, chinless face, Slim was a natural comic. The audience was bound to be disappointed if I appeared in place of the great Slim Pickens. Rodeo would require a full-scale investment of time and money, and there would be little room for other interests. I knew that I wasn't ready for big-time rodeo responsibilities.

Slim found other clowns to take his contracts, and so the rodeos went on without me. The champions made money; lesser cowboys spent their meager winnings on travel, doctor bills, and entrance fees, and soon disappeared like me from the rodeo scene.

Looking back, I was lucky to quit when I was in one piece. Through the years the rough stock got rougher and the bulls more savvy and dangerous. Ranchers began breeding superior bucking bulls and pairing talented bucking mares with talented bucking stallions. For a contestant, the margin between winning and coming up empty was often measured in split seconds, and how long you could take that beating and keep your body healthy.

More and more bull riders were tying themselves on the beasts using what became known as the suicide wrap and were often hung up, bucked off with their hands still caught in their riggings. I was a cape fighter, and the cape would only be in the way of my getting a bull rider's hand out of his bull rope fast enough to save him.

I suddenly found myself with only two real choices, paying money to sit in the stands and watch or, what was easier, just staying away. I loved the ranch, and wanted to become a writer and a family man. The friends I had in rodeo would always be my friends, and maybe I could become better at other jobs than I had ever been in the rodeo arena, where I'd had my moments but was never great. I could look back on some real adventures, like coming out of a chute on such great horses as Coburn's Cheyenne or Harry Rowell's Sontag, producing and starring in the first American rodeos in Arles, France, fighting bulls and clowning with Slim in San Francisco's Cow Palace, and winning honors with my rodeo photography.

I moved back to the ranch for good. Maybe it wasn't the Yamsi I loved back before the war, but it was a good life, and I still had lots of friends there and things to write about. Like Toy Brown, one of my Indian friends from Beatty.

Toy was a huge man with a head the size of a basketball, and he could have stolen his gap-toothed grin from a jack-o'-lantern. He worked behind the bucking chutes at many a rodeo, and producers like Mac Barbour claimed they couldn't put on a rodeo without him. Toy was a master at tightening the flank straps on bucking stock and could make you get bucked off or let you ride and maybe win.

Properly tightened, not too tight but not too loose, a flank strap makes an animal kick back. Toy knew his rough stock and how to get the best performance out of every one of them. When a cowboy got bucked off and staggered back to the chute, Toy Brown's ear-to-ear grin was often the first thing he saw.

The Browns were a ranching family who had a good cattle ranch just east of the Beatty, Oregon, rodeo arena, and Toy had a reputation for being a good man with horses and a steady hand on a haying crew. He seemed always glad to see me, and I counted him as being a good friend.

One day as I was driving through Beatty in my pickup, I saw Toy sitting on the front porch of the general store and stopped to visit. He was holding his great moon face in his brown hands, and for the first time ever, he failed to grin at me as I approached.

"What's the matter, friend Toy?" I asked. "You look down in the dumps."

The big Indian motioned me down beside him. In the distance on the hillside we could see the ranch house where old Toy had been born.

"You've known me a long time," Toy said. "I always worked hard, didn't I?"

I nodded, wondering where this was headed.

"I saved my money and made something of the old family ranch, didn't I? Never bothered anybody or raised hell like some of my neighbors. I married a woman with a grown daughter, and I was happy, and worked hard to give them a good life. Well, them two went through all the money I had, even cost me my ranch. This morning the daughter told me she was going to pack Mama's bags and take her away. You know what I told her?"

"What, Toy?" I asked.

"I told her if she tried to take Mama away from me, I was going to shoot them both. That's what I told her."

Poor, kind, gentle, old Toy. He tried to light a cigarette, but his hands were shaking so much most of the pack slid out and fell in the dust.

I struck a big kitchen match on my thumbnail, and when I lit his cigarette for him, I was surprised to see he had tears in his eyes. He lumbered to his feet and went off toward his pickup without even saying good-bye.

I was at the ranch the next day eating lunch in the kitchen when the phone on the wall rang my ring. It was Toy. He was in the Klamath Falls jail and wanted me to bring him a carton of smokes. He had done just what he had promised. When the girl tried to take her mother away from home, he shot them both.

Toy had flanked his last bucking horse. For several years he was a trusty at the Oregon State Penitentiary, and grew flowers in the gardens. His hair grew white with age, and he could have been granted a parole anytime he wanted, but he stayed on to the very end. Knowing Toy, I'm sure there was never much sense of remorse over what he had done. He had warned the women ahead of time, and that was that.

Soon after Toy left the rodeo scene, I lost another old friend, an aging Indian saddle bronc rider named Jerry Choctoot. Jerry was an institution at Mac Barbour's rodeos. More often that not, Jerry would win some money in the saddle bronc riding the first day of a rodeo, but the other cowboys would make sure he was too drunk to ride the next day.

I was in the arena that year in Klamath Falls, Oregon, when Jerry showed up drunk to ride his second horse, which was already saddled and ready in the chute. I cornered Mac Barbour behind the chutes and told him that Jerry was in no condition to ride, and that I was going to unsaddle his horse and turn him out. Mac seemed to agree and went over to talk to him, but suddenly I saw Jerry crawling down on his horse, and Mac Barbour pulling the pin to let him out into the arena. One moment, the horse was bucking hard; the next, Jerry Choctoot had ridden his last bronc and lay dead of a broken neck. The rodeo went on, of course, but that was the last Mac Barbour show I ever went to.

During the long winters at Yamsi, I had plenty of time to do what I had long dreamed of doing, and that was to write books about wildlife and the West. I was wandering through northern Nevada one day, looking for something in the way of history that I could write about, when I stumbled upon the story of an Indian family who had fled the reservation system and gone back to living wild and free. Whenever I could shake myself loose from the ranch, I would rent a room in Winnemucca, Nevada, and spend a few days in the local library, reading anything I could find in old newspapers about the family. Oddly enough, there was plenty to read.

The head of the family was called Shoshone Mike, and he was married to Snake, a Ute woman from northern Utah. For many years after he and his family fled the Fort Hall Indian Reservation in the 1890s, they wintered south of Hansen, Idaho, in the mountains just above where Rock Creek flows through the Charlotte Crockett ranch.

Although Mike and his family disappeared for long periods as they wandered in search of food, they would always return to their winter lodge along Rock Creek. The Crockett family knew them well. Old Mike taught the Crockett boys how to hunt with bows and arrows, and the white and Indian

Chapter Twenty

WHEN THE ZX QUIT USING DRAFT HORSES and went to tractors, they had some fine animals that would have been slated for slaughter had I not bought the lot of them and moved them to Yamsi. There were some spoiled horses in the bunch, but I was determined to break them of their bad habits. I fed cattle with them, logged with them, and hooked them up to wagons to haul fence material. Whenever a team would get to be trustworthy, I would peddle it and start some of the others.

The market for teams wasn't very good. Since the war, most ranchers and farmers had gone to tractors. The old teamsters by now were a thing of the past, and it was hard to find a man who knew how to harness a team and drive it. But there were still a few ranchers around who were sentimental about old ways, and I sold quite a few good teams through my enthusiasm. One little team of Shires which I unloaded on a stranger ended up being driven by Amanda Blake, Miss Kitty, on the TV series *Gunsmoke.*

I missed talking rodeo talk, but often some of the old gang would drop by to visit. Every so often, I would be awakened before daylight by shouting in front of the Yamsi house, and there on the lawn I'd see Slim Pickens on his way to some movie location or another, often long-haired and bearded for a current role. We would spend a few days on the ranch, fishing and driving teams, while he studied lines or rested up from a movie.

One fall, I had caught up a fine young bay team and broke them to lead. We were busy getting a bunch of steers ready to ship to market, but I got up early and harnessed the team for the first time and tied them in the corral to let them get used to the harness. They were wild and woofy; the discipline of being tied to the fence would do them some good.

With my crew of hungover cowboys, I spent the morning gathering yearling cattle in the fields, then held the bunch against a fence while we cut out inferior steers and shaped up the herd for market. The days of shipping cattle by rail were long over. Cattle trucks were due to arrive that night at the ranch corrals, so I was under real pressure to get the work done.

What we didn't realize was that Slim had arrived at the ranch. Finding the place deserted, he went over to the corrals and spotted that young team tied to the fence. Somehow he managed to get them hitched to a wagon. We had almost finished our day's work when they came stampeding down through the field, panicking cattle as they ran. Slim had all he could do just to stay on the wagon. "Whoopee!" Slim hollered as the team ran through the bunch we were trying to hold, scattering animals across the meadows.

Everywhere I looked, there were steers running and saddle horses bucking.

"Hey, kid!" Slim shouted as he almost ran me down. "Why didn't you teach these sons of bitches to whoa?"

I expected the cowboys to get mad and quit, but when Slim finally got the runaway team stopped and the cowboys recognized that great hulking figure as their favorite rodeo and movie star, all was forgiven. Slim had an unusual ability to make instant friends of the strangers around him. His big grin always put everyone at ease. When Slim took a shine to a person, he made him feel as though the guy was his best friend in the world.

The big cowboy always arrived at the ranch without warning. He had been making a pilot film on the Oregon coast for *The American Sportsman* when bad weather canceled the shooting. Slim arrived at the ranch with the crew and commenced shooting where they had left off.

We had spent a week on camera, trout fishing on the river and reminiscing about old rodeo adventures, when Slim began to complain about headaches, and got ready to head back to California to see a doctor. He must have sensed that the diagnosis would be serious. As I drove him in to Klamath Falls to the airport, he opened up to me as he never had before. He had never mentioned my saving his life in the Cow Palace by pulling the Rowell bull Twenty Nine off his body so many years before. Now he thanked me for the extra years I had given him.

He phoned me one evening full of hope. He had been diagnosed with a brain tumor but had found a doctor willing to operate. I never heard his voice again. Slim lived his life at the edge of danger. The final irony was that the man who lived with violence in real life and rode the A-bomb in the movie *Dr. Strangelove* died of something he couldn't control.

Slim's memorial was held at his and Margaret's beautiful home in the Sierras of California. It was an awakening for me. I sat looking about the congregation at rodeo stars I had known for years and was struck by how they had all aged. That night, when most of the folks who came to honor Slim had gone home, a few of us gathered in a small basement room and talked about old times. Rex Allen took out his guitar and sang some of the western songs Slim had loved, while rodeo announcer and master storyteller Mel Lambert told story after story about his long friendship with Slim. The party went on long after midnight, until Margaret heard noises in her basement and, assuming that her guests had long ago departed, came down to investigate.

During Slim's life, one of his best pals had been Mel Lambert, and after the funeral Mel adopted me. As a storyteller Mel was able to switch from one voice and character to another at will. Not long after I got home from Slim's funeral, I got a strange telephone call from an Italian truck driver who claimed, in fractured English, that he had a truckload of salmon from the Oregon coast he was supposed to deliver to me as a gift from a friend. He was having trouble finding the ranch, and the fish were beginning to spoil. He had already been kicked out of his motel because of the smell, and was desperate to find me and dump the load. I became just as desperate to head him off and tell him I didn't need a load of rotting fish. He yelled at me in broken English, and I yelled back.

Finally, I felt so sorry for the poor truck driver that I was working on a solution when I heard a chuckle at the other end of the line. The truck driver was Mel Lambert.

Mel and I found we had a lot in common. He had grown up in the little town of Chiloquin, the biggest town on the Klamath Indian Reservation. Part Umatilla Indian from northern Oregon, he was a navy pilot during the war and never lost his love of flying. He did voices for the movie industry, and was a superb humorist who was often described as "the funniest man in Hollywood." One movie part he did was that of the harbormaster in *One Flew Over the Cuckoo's Nest*. Another was a role in *Three Warriors*. His career as a rodeo announcer spanned fifty years, and Mel is in the Rodeo Cowboy Hall of Fame, along with Slim Pickens and Montie Montana.

Mel ran a used car and airplane outfit in Salem, Oregon, and was never happier than when he was in the midst of a trade. One of my cherished possessions is a turquoise ring Mel gave me. Well, he didn't exactly give it to me. I was staying with Mel and his wife, Pauline, in their ridgetop home, when Mel brought out a big tray of rings he had traded for from an Indian he had befriended in Arizona, and told me to pick out a ring as a gift.

I took one look at the tray and knew that, whatever the deal had been, the Indian had come out way ahead. They were the kind of dyed turquoise rings you didn't want to wear in the bathtub for fear your body would turn blue.

Not wanting to hurt Mel's feelings, I pretended to agonize over my choice. Right in the middle of the tray of a hundred rings was a gorgeous one of cobwebbed turquoise, a museum specimen. "I'll take that one," I said, slipping it quickly on my finger. Moments later I heard him in another part of the house berating Pauline for putting his best ring back in the wrong tray.

Mel, Montie Montana, and I were in the Rapid City, South Dakota, airport one day, waiting together for a plane. Montie had once roped President Eisenhower as part of his famous trick-roping act and had starred in many a western film. As usual, he was immaculately dressed and superbly handsome, while Mel looked as though he had just escaped from the local jail.

It wasn't long before Montie's fans discovered him, and he was surrounded by fifty or so strangers asking for autographs. No one cast a glance at poor Mel.

Finally a man came up to Mel and said, "And who are you? Should I be getting your autograph too?"

Screwing up his face and combing his hair sideways, Mel suddenly became the popular comedian Red Skelton, and launched into a Clem Kadiddlehopper skit.

"Oh, my God," the man shouted. "It's Red Skelton!"

Montie's fans deserted him and thronged around Mel, who blithely began signing Red Skelton's name.

Mel could always be counted on to help other Indians in trouble. In appreciation, the chief of the Klamath tribe made him a present of a huge ceremonial tipi, and sent some of the tribe to set up the structure on Mel's ranch property overlooking the city of Salem, where it could be seen by everyone traveling the freeway.

The tipi was the apple of Mel's eye until one day, as he was driving by on his way to his office at the Salem airport, he glanced over just in time to see a dozen big Hereford steers disappearing into the tipi door. Steers being volatile, there was no way he could get the animals back out the entrance. Each of the frightened animals made a different hole getting away, and there wasn't much left standing.

One of Mel's pilot friends was Hoyt Culp, warden of the Oregon State Penitentiary. Through the years, Mel worked with Hoyt to get a lot of Indian parolees back on their feet, and he was as revered by Native Americans as he was by rodeo cowboys. His house was often full of Indian children from the Warm Springs Reservation, whose families Mel and his wife had befriended.

One little girl turned out to be a holy terror, and since Christmas was a week away, Mel figured she would want to be back with her family on the reservation. The girl was pretty angry, but Mel bundled her up anyway, and off they went. He had just sold her old man a brand-new Chrysler and figured by delivering the girl, he was doing her father a favor.

He was driving across the reservation when suddenly her father passed him going the other way, driving the new car.

Mel slammed on his brakes. "There goes your dad in his big Chrysler!" he exclaimed.

"That's not my dad!" the girl snapped. "And that's not his car."

Mel did a U-turn and pretty soon caught up with the old man and flagged him down. He got out and walked over to the Chrysler. "I brought your daughter back," he said.

"Oh," the man said, not looking too happy. "Couldn't you have kept her roun' till after Christmas?"

Mel looked through the windows of the new car. There was blood all over the seats, the headliner was ripped out, and the upholstery was demolished. Several windows were busted out, and the seats were littered with shards of glass.

"What on earth happened to your new Chrysler?" Mel asked.

"Well," the Indian said, "I was drivin' down the road, me an' my frien', an' we saw a big buck deer standin' in a fiel' an' I took my rifle an' shot him. We loaded him in the backseat an' I was drivin' back to my place 'bout ninety miles an hour when the goddam ting come to an' tried to climb in the front seat with us. I couldn't stop so I just grabbed my rifle, stuck it over my shoulder, an' started pullin' the trigger. That deer, he didn't like to be shot at I guess and tried to kill us with his horns. Finally, my last bullet got 'im right in the throat. Say, my old lady don't like the color of this car you sold me. Maybe I bring it back an' you give me 'nother one."

Chapter Twenty-One

NOWDAYS, WITH A NORMAL HUMAN LIFE SPAN, an old cowboy like me will have said good-bye to about three of his favorite saddle horses or about five of his favorite cowdogs. One of the hardest things for a rodeo cowboy to accept is knowing that one day his body will give up, and that someday he will have to pay for a ticket in the stands instead of having free access to the arena and a seat along the chutes. It's tough sitting there watching from the stands, because you mentally ride each bronc out of the chutes and fight each bull, and your muscles twitch and strain as though you are actually doing it. To compound the problem, you have to watch the new age of cowboys doing things differently, and with a different style. The modern clowns tend to dart around, making fools of the bulls, whereas Slim Pickens and other cape fighters tried to make the bull look good. To be fair, I have to admit that modern cowboys are more talented and the stock rougher than any I ever saw in my youth.

From time to time through the years, I would take my old Mexican bullfighting cape out of the closet and practice on the lawn, noticing how much heavier the cloth seemed from the days I used it in the arena. I would stand there doing veronicas, sweeping imaginary bulls past my body, a little sad that I wasn't headed for a rodeo somewhere, and wishing often that I had done a better job on the last tough Brahma I fought in California, so many years ago.

That last awful bull! He had put me in a cast, that one. On his very first charge he'd jerked the cape right out of my hands with his horns, and when I tried to retrieve the cape off the ground, he'd thrown me like a half-filled sack of spuds right over the arena fence. Had it not been for that barrier, he might have followed through and finished me off.

Night after night, in the months following my injury, I dreamed about that animal, fighting him time and again, trying to figure out where my timing had failed me. Proud as I was, I knew I had to face him again, and by now that bull had been to several more rodeos and learned more tricks. I was just out of the cast and still weak when I got a chance to fight the same animal that hurt me at a rodeo in Petaluma, California.

I could hardly hold up my cape, and I didn't do much more than protect the fallen rider and survive the animal's charges. I swear that animal remembered me. Nervous sweat poured down my back, and I realized right there that, after that performance was done, I was going to have to quit rodeoing. The joy had gone out of something that had always been fun for me.

For the next forty years that bull was often in my dreams. I was busy ranching and writing books, my rodeo days were long past, but sometimes I would wake up in the night all covered with sweat from fighting that animal. What really bothered me was that I'd never get another chance to relive that afternoon. Why hadn't I tried harder and retired on a positive note? Worse yet, as the years passed, I began to wonder if I'd failed because I was scared.

On my sixty-fifth birthday, I happened to be at a rodeo in Portland, Oregon, and was headed for a seat in the stands when I heard someone call my name. It was the old rodeo clown, Mac Berry, who had been a friend of Slim Pickens and had watched me clown with Slim years before. It felt pretty good to be remembered.

Mac grabbed me by the arm and pulled me into the arena. "Have I got plans for you!" he exclaimed. He pointed to a pen on the hillside behind the bucking chutes. "You see that big Brahma bull? He's gentle as a kitten. Why, kids can climb all over him. I'll have Bob Tallman, the announcer, tell the crowd you are coming out of retirement after forty years to fight this two-thousand-pound bull. At your age, everyone in the crowd will expect disaster. I'll turn the bull out of the chute, and he'll walk right up to you. All you have to do is reach out and pet him, and the animal will expect grain and follow you around the arena. The crowd will laugh themselves sick."

"At my age, I'm pretty happy just sitting in the stands," I said.

"Please!" Mac pleaded. "For old time's sake. I need to rev up my clown act, and the crowd will get a real chuckle out of this. Do it for Slim."

I wasn't much dressed to rodeo, but I was tempted. I spotted Mel Lambert along the fence and handed him my wallet, telling him I was going to exhibition a fighting bull. He stared at me and moments later handed me a slip of paper on which he had scrawled:

> To whom it may concern. I, Dayton Hyde, hereby bequeath my D. E. Walker saddle, my black sports car, my turquoise ring, and all my earthly possessions to my good friend, Mel Lambert of Salem, Oregon.

"Sign this," he said. Beneath his levity, I could see he was really angry with me. "You goddam fool!" he snapped suddenly. "The cemeteries are full of cowboys that tried to come out of retirement and rodeo just one more time."

I wanted to tell Mel that it was all an act, that the bull was gentle, but suddenly there was no more time. Bob Tallman, the rodeo announcer, was out there in the arena with his horse and microphone, giving me a big buildup. Vaguely, I heard him say, "The last time Hyde fought a bull was before I was born. It was at the San Francisco Cow Palace over forty years ago!"

I took off my blue suede sports jacket to use as a cape and stood in front of the chutes, just as I had stood ready so many times, so many years ago.

There was a big uproar behind the chutes as Mac tried to move the big gentle bull through a corral full of rodeo Brahmas, and cowboys had to use electric cattle prods to break up the fight. I yawned as though bored. I could see friends in the audience staring at me, wondering if I had gone out of my mind.

I believe to this day that Mac Berry's plan would have worked had not they used an electric cattle prod on my bull. The chute gate opened, and out roared that big gray Brahma, looking for someone to kill. Instead of charging at me, however, the bull rushed past, leaving me standing stupid and alone in front of the chutes. Instead of a belly laugh from the crowd, there was only a faint, nervous titter of embarrassment. I wanted to crawl under a cow flop and die.

The raging animal put the cowboys out of the arena and came back to center stage, hunting a human body. Usually a flank strap, tightened just in front of a bull's hips, slows him down and makes him easier for a clown to handle. Mac had sent this bull out without a flank, and a chill went up my spine.

I could have made it to the fence, but I seemed to hear Slim Pickens's voice out of the past. "Take him, kid! You can handle him fine! Pretend you're the greatest bullfighter in the world, an' you'll do great."

The bull's eyes focused on me, his head shook in anger, and he flung up his tail and charged. He was almost upon me when I knew that I was in real trouble. My feet were roots grown down into the arena floor, and my reflexes were all gone. I was going to die! I felt a hand at my back as though Slim were pushing me on to disaster, keeping me from fleeing in terror.

Two strides away, the bull dropped his head to catch me. He was coming like a freight train, and my feet still wouldn't move. He could see my feet under my jacket cape and knew exactly where to hit. Suddenly, it was no longer me out there waiting to die. I was an actor playing the part of a great bullfighter. Instead of trying to sidestep, I held fast. Then, as the massive head hit my cape, I stepped toward the bull's shoulder.

I felt his smooth skin brush my belly, and he was past. By the time the second charge came, I was enjoying my role. I felt the moist breath of the animal on my face as he rushed past. This time the bull hooked at me, and I sensed I had pushed my luck too far. The bull had me figured now, and the next charge would be my last. In the old days I might have led the animal through a series of a dozen charges. Now I knew that I should quit while I was ahead.

As the bull stopped and stood shaking his head angrily, I turned my back and walked away from him. The crowd gasped, thinking maybe how brave I was, but I was watching the face of a woman in the audience. It was an old trick Slim had taught me. Had her face suddenly tensed with fear, I would have left slippery tracks getting the hell out of there.

Bob Tallman, the announcer, rode up to me on his horse. "Where on earth," he said, "when and where on earth did you learn to fight Brahmas like that?"

"When I was young," I said, ripping out what was left of the lining on my sport coat. "When I was young, a long, long time ago."

As I headed across the arena, I was lost back in the past. I was at a rodeo somewhere, riding Blackhawk through one helluva storm. I could hear the crowd cheering and sensed that I was within seconds of making a qualified ride. In that last instant, as a tribute to that great horse, I threw my bucking rein into the air and stepped hard in my right stirrup.

Sailing through the air, I landed on my feet and watched as Blackhawk found the gate to the catch pens and vanished from sight as my dream faded and was gone.

Mel Lambert met me at the gate as I left the arena. He looked grim and shook his head in disgust, but I knew that he was just a little bit proud. "Damn," he said. "I was sure hopin' to get back my turquoise ring."

Chapter Twenty-Two

SOMETIMES ON LIFE'S LONG ROAD come sudden turnings. In 1988 I was buying cattle in Nevada to pasture at Yamsi in Oregon, when I passed government-sponsored holding facilities near Lovelock, crowded with hundreds of captured wild horses. I stood outside the fence and saw the hurt and dejection in their eyes. They had been born to the wild and were meant to be free. I had grown up with wild horses on the ranges surrounding Yamsi. I had a feeling that, somehow, I was meant to help the animals escape.

I had an impulse to sneak into the feedlots in the dead of night and open all the gates, but I dismissed the idea as wildly impractical. The horses had been gathered by the Bureau of Land Management in the first place because their numbers on the range had exceeded their food supply. Releasing them wouldn't solve the problem and would only add to their misery.

Instead, once I had shipped cattle to the ranch, I headed off to find a place in the West where there was land enough and grass enough to set up a wild horse sanctuary so that the unadoptable wild horses languishing in feedlots could run wild and free. I had little or no money but lots of enthusiasm, and I hoped there would be some good, caring people somewhere who would embrace my dream and help.

For a time, I awoke every morning in a different motel, hoping to find one with man-sized soap, meditating silently with airline stews in early-morning airport limousines, wishing my mail — *and* my luggage — would catch up with me, hoping daily that fate wouldn't deal me a talkative seat companion, or another large lady like the one who reached over when I was dozing and ate my lunch right off my tray.

I had sworn that wild horses couldn't drag me away from Yamsi, but they ended up doing just that. I turned the ranch over to a manager and took off to pursue my dream. All too suddenly, I left my friends and livelihood behind to get those wild horses out of the dust, disease, and boredom of the feedlots, and let them run free on better land than they had seen this century. I found that range in South Dakota.

Why me? Because I owed wild horses something! I have one ache and one pain in my body for every horse I ever met, but so many memories and so much joy. I was maybe fourteen when I rode my horse up over a lava rimrock and surprised a wild horse family dozing on a spring meadow near Fuego Mountain south of Yamsi. They were all grullas, blue velvet mares and a blue velvet stallion, all with black dorsal stripes. One moment their hooves rang bell-like on the lava table rock, then the next they had vanished into lodgepole thickets, and only the snapping of branches and a pale shroud of pumice dust marked their passing. For years I rode those horses in my dreams.

I spent the next six months in Washington, D.C., talking to every legislator who would listen. The Wild Free-Roaming Horses and Burros Act of 1971 had brought about a population increase beyond the ability of federal lands to support it. Tired of problems with excess wild horses, Congress finally embraced my plan, and instructed the Bureau of Land Management to work with me. I tied up one of the best horse ranges in the West, an eleven-thousand-acre tract of land at the edge of the Black Hills in South Dakota. Soon trucks were rolling in daily, bringing in captured mustangs, most of which had no idea of fences or that they couldn't get back to their home range simply by galloping west. There was one old mare I'll never forget, perhaps because she became a symbol of what brought me here.

The Appaloosa mare came crashing out of the livestock truck, pounded down the loading chute, tried to climb the corral fence, then charged me, yellowed teeth bared, ready to eat the world. As I scrambled over the top rail, she ripped the back pocket off my Levi's, then whirled to kick, blasting a jag of splinters out of the dry juniper logs. The old mustang was shaking with stress, and so bug-eyed with rage that it seemed inconceivable to me that one day we would be friends.

Her one ear chewed off, maybe by her mother in birthing, her odd color, the pink warts on her nose, and an ancient bullet slash across the flat of her right hip made me guess where I'd seen that animal before. She'd maybe come halfway across America from Little High Rock Canyon to haunt me for something that wasn't my doing. In northern

Nevada, twenty years back in time, some twelve hundred miles away.

I remember that day. I had borrowed a saddle horse from Butch Powers's outfit,* skirted the Black Rock Desert by night, eased my horse to the sun-baked rim of a canyon, and sat looking down at the shine of water along the bottom, wondering if somewhere amongst that chaos of ancient, fractured rock, there might be an animal trail leading down to the cool.

Even the tiny fans of aspen leaves hung silent, as though painted on a museum tapestry. A rock wren piped from some shaded bower amongst the rims; a lone buzzard patrolled its desolate kingdom; a Townsend's solitaire began its song from a stunted juniper, then gave up, as though singing in that heat were too much trouble.

The band of wild horses came tiptoeing off the far rimrocks, working its silent way to water. An old black lead mare came first, stopping frequently to stare and work the canyons for the scent of danger. Behind her were three dry mares, equally watchful, then a mare and foal, followed by a battle-scarred stallion. The foal caught my eye: an Appaloosa, one ear cropped, rump oddly splattered with white as though a magpie had roosted there, the rest of the animal blue like sage smoke over a November campfire.

The little filly stuck to its mother's side like a bad smell, even though there was only room on the trail for one. Dislodged by time, a small rock rattled and clicked down the

* Butch Powers was then lieutenant governor of California.

hillside, then chunked into the sun-dappled water on the canyon floor. The wild horses froze, thinking about that rock. The Appaloosa's mother restrained the foal with her nose, and for moments, the band seemed sculpted in stone.

Suddenly, the willows along the bottom erupted in rifle fire. The stallion humped up, leaped, screamed once, and died. I shouted at the mustangs to run, but only the mare and foal made it up over the narrow trail to safety. I caught one last glimpse of the little Appaloosa as it dragged with one crippled hip after its mother, then they were lost in the sun. I roared a torrent of bad words at the horse hunters and got the hell out of there, never dreaming that I would see that filly again.

The Bureau of Land Management records on the old Appaloosa mare are slender indeed. Appaloosa. Sex, female. Captured, 1987, on Black Rock Desert, northern Nevada. Shipped by truck to the Black Hills Wild Horse Sanctuary, Hot Springs, S.D., from holding facilities at Bloomfield, Nebraska. Released, September 21, 1988.

I didn't see much of the old mare for a while after I opened the gate to freedom. Besides the Black Hills unit, I had two other sanctuaries to manage with 1,800 wild horses, and the old App stayed up high, where she could see Wyoming to the west, as close to Nevada and her old range as she could get.

She finally picked up with a Kiger mustang off the Steens Mountain range in Oregon, and I knew I had a chance to settle her down, since her new friend was addicted to the grain I kept in the back of my pickup truck. Soon the old App would follow the other mare down off the heights to fight over the pile of oats I left as a peace offering. Once she even nickered to me, but I was careful not to let her see me smile.

There is among wild horses an innate sense of home. Horse bands, unfortunately, tend to be too sedentary for their own good and the health of the range they occupy. When nature dries up the water holes and scorches the grass, wild horses often fail to move on. As the land dies, the horses die with it. Years back, predators such as cougars, wolves, and humans harassed the horses and drove them to new ranges, mixing up the herds so that far less inbreeding occurred.

The grasses on the sanctuary were shaped by thousands of years of prairie fires and buffalo grazing. The bison gnawed the grass to the nub but moved on, migrating to new pastures, giving the plants a chance to grow strong again and go to seed before the next invasion by a herd. The grasses used prairie fires to keep pine trees in their place. The light accumulation of dead grass was soon taken care of by nature, in the shape of cool fires, insects, or grazing.

My days on the sanctuary were soon spent playing wolf or Indian, moving the horse herd on to fresh ranges even though the horses would pace the dividing fences for days after each move, trying to return to where they had grazed before.

The mustangs had never forgotten the trauma of their capture and would get silly anytime a man a-horseback or helicopter appeared in the distance. But they would allow me to drift them afoot, or bump along behind them in my well-used pickup truck. Horses can only concentrate on one thing at a time, and I soon learned that when they were headed in the direction I wanted them to go, it was best not to distract them.

Wild horses live in a matriarchy. Each band has a lead mare who determines where to graze, when to leave for water, and how best to elude the enemy. The stallion stumbles along behind, pretending importance, mainly concerned with keeping his harem together so that no rival stallion can steal a mare. When danger threatens, the stallion trails the band, keeping between his mares and the enemy.

The groups of wild horses on the sanctuaries had been captured in Oregon, Nevada, and Wyoming, and contained a good many old mares that had once enjoyed status in their herds as lead mares. They did not give up their old habits easily.

Whenever I would attempt to drift a gathering of horses on to a different part of their range, every old lead mare would remember her past, and dash through the herd in an attempt to gather a band and lead them to safety. My herd would suddenly fragment into six different bunches, all leading at a long trot for rocky steeps where they knew I couldn't drive my pickup.

The old App, of course, was the worst. She would run through the herd with such conviction and semblance of real terror that the others would assume she had seen some monstrous danger, and away they would race after her, the old mare running flat out and sassy, thundering down trails only she knew about. In time they would do a big circle and end up back where we had started.

The mare and I were two generals, each with opposing

battle plans. Time and again she defeated me, leaving me with a day wasted and my pickup truck severely damaged from racing forty miles an hour over a devil's garden of ancient rocks. There came a time when there was no turning her from her headlong flight. Her wispy tail would come up, her eyes would glass over just as soon as I tried to move her band, and off she would race, charging past my vehicle as though it were invisible.

It was grain that was her final undoing. Every time she sprinted off with her band, I fed the remaining horses a taste of sweet chop, containing a mixture of oats, alfalfa, molasses, and corn. It didn't take long to erode her following. The next time she went running off in pretended terror, the other mares stayed behind, and soon she was the first mustang off the mountain to get the grain.

Little by little, the wild horses left their fears behind and accepted my presence. Often as I wandered afoot in the darkness, I could hear them near me; a flinty hoof striking rock perhaps, sounding a blunt bell. Often, far down a ridge, a lonely mare nickered for a friend separated, somehow, by the night's grass-to-grass wanderings, and maybe a black shape nearest me would answer back. From the crest of a hill, a mustang might cough the night dust from aging lungs, and I'd hear the snuffling downslope as animals grew restless with impending dawn and inevitable change. Obscurity of vision for full sight; cold starlight for warm sun; wet grass for dry; night moths for daytime biters. In the dark, the horses were black on black, invisible yet surrounded by a field of warmth and energy that kept me from colliding. A soft wall warning me that I was too close.

In any dawn, the moment comes suddenly when the white horses in the herd first take shape, like scattered blobs of remnant moonlight, luminescent pearls on the hillside that vanish intermittently as dark shapes drift between them and me. Palominos and light buckskins, sorrels and roans, then bays and blacks. The herds themselves take form in the pinkish dawn. I can almost read horse thoughts as grazing buddies come together along the edge of the herd and drift away like human couples leaving a bar at midnight, hoping others won't notice and follow.

The old App rises from where she has been lying down, stretches the stiffness from ancient limbs, and picks her way down through the rocks to where I have put out grain. In a horse herd there is no respect for age. Younger horses have followed my every movement and dash in to fight each other. The old mare goes back to her ridge unfed.

But the old lady figures it out. The next morning and the next she anticipates my route and meets me in a valley screened from the others where she can have her own ration of grain and dine in peace.

Seasons slide into each other, and the wild horses and I grow gracefully old together. The first sparkle of an early frost comes suddenly to this bit of prairie, each twinkling star mirrored on its own blade of grass. Daylight will come a bit later this morning, rubbing its eyes like a bear not quite ready for its winter sleep.

There is not much fun and games in old mustang mares. During the summer they have given all their extra energy to provide milk for demanding foals who have clung to their sides like shadows on a sunny day. In vain the foals have tried to tease the mares into playing games. Now the youngsters begin to socialize. They race each other across the prairie, the rocky, arid ground beating hardness into tiny hooves, leaping gullies, prancing, dancing, pawing over backs in mock matings, acting silly, tails high, fair to bursting with energy like children high on sugar.

Playing hard, they chew on each other's manes, rising together in a juvenile waltz, cavorting like quarreling stallions, biting at forelegs, backing into each other to drum a tattoo of hoofbeats on the other's ribs.

The insects hide from the night frosts. The grasshoppers, which have sucked the grasses dry, seek out hidden crannies against the cold air which pools in hollows along the valley floor. A few bumbling botflies, playing innocent, meander their slow, clumsy way around the horses' bellies to lay eggs on the hair of drowsing mustangs, then are gone until spring.

In November, shortly after midnight, a group of wild horses comes down from the rimrocks, slipping in so quietly I do not hear them come. I look out my window and see the old App, ghoulish in the moonlight, standing watch as the others move through piney moonshadows to drink.

Gone from their coats is the sleekness of summer. Already these lovers of the storm are jacketed for winter. The first snows will pile on them unmelting, until blacks, bays, paints, roans, grullas, and sorrels will vanish, white on white. There is so much shelter here of pine and rock, but give the mustangs a good blizzard, and they seek out the highest hills, standing just below the crest, turning tail to the buffeting winds, each taking its turn at the windy edge of the herd.

Winter always seems to start somewhere else. It comes in from someone else's range, and suddenly, most often in the night, it lays a white blanket over the sanctuary, obscuring the golden carpet of leaves beneath each cottonwood tree and leaving each wild horse to graze in the tracks of the one which walks before.

There are no flies in winter to plague the mustangs, and each horse's tail hangs like a plumb bob, at rest until spring, except to register, perhaps, a swish of anger when another animal ventures too close.

I take the old App a big, round bale of hay and place it for her where others cannot see. For a time she watches from a distant hillside, then curiosity brings her close. Lured by a tantalizing odor of summer, she bites at the bale and jerks out a long beard of grass. In her lifetime, grass has never behaved thus, and she whirls in terror. Unable to open her mouth and release the grass, she thunders away, bucking and striking, until at last the beard falls to the ground. She stops at a distance and gradually works her way back to the fallen clump, sniffing it carefully. Now it is in a more familiar form, and she finishes every wisp.

Soon she is back at the bale, willing to try again. Another wisp of hay, and she whirls off, but the distance is shorter now. Some thirty feet out, she drops the hay, then eats it on the ground. Later that afternoon, I pass that way again, and she is there still, head buried in the bale as though she has been raised on such easy fare. Winters are long and hard. The old App saves her energy and makes no motions not necessary to her survival.

Spring peers around the corner, smiles, then retreats. Like domestic horses, the mustangs are quick to sense meaningful changes in the weather, galloping wild and free down mountainsides, across gullies, splashing noisily through meltwater rivers, hooves making the first thunder of the season across the hollow drum of the plains.

Traveling, a bunch of wild horses generates its own music. There is not only the drumming of hooves but whinnying, of foals for their mothers, mothers for foals, and one friend for another. There are softer sounds too, the cough from a dusty lung, or the contented snuffling horses make when life is good.

This spring, the pasqueflowers turn the rocky hillsides along the Cheyenne River to purple. Most of the wild horses move high to the sandy ridges, where the blackroot grows. Ancient Indians led their winter-weak horses to these same ridges, and in three weeks they were ready for the warpath.

The old App stays behind. Once or twice she ventures to the edge of the spring-swollen river and looks across with longing, as though to follow the others to the high places she has loved ever since she came here. She waits for me every morning, listening for the rattle and squeak of my truck as I head off cross-country through the pines to give her grain. Her nostrils flare as she approaches, and she snorts like the blast of an old musket, unwilling to dispense her friendship lightly. I would feel better if her attraction to me were not bought with grain. She tiptoes forward, then whirls and trots down the hillside. It is an act she, out of pride, must put on every morning.

Her tail is a disgrace. Never long and full, it is now wispy like her mane, but nevertheless clogged with burrs of wild licorice and burdock. Flies cluster on her back, just out of reach of her club. She is so close now I can see every nose wart, every blemish on her hide. The ancient bullet crease across the top of her hip has formed a ridge of hard horn, as though some excess hoof material has oozed up like lava through a fissure in her body. Her stub of an ear makes her look somehow out of balance.

There is meaning to her aloneness. I can read the signs. The old mare hardly touches her grain, and she stands listless, eyes sunken. She won't be here tomorrow. She has already picked her spot to die, beneath a gnarled old juniper overlooking the range she came to love. She will spend her last hours there standing alone. At best, my ministrations of grain have brought her a couple of extra years of the good life. Tomorrow she will be part of the eagles that fly over the canyons.

High on the ridges above the Cheyenne River, I see wild horses running in pure joy. Life goes on; my job as a volunteer goes on. Since that day I left Yamsi, I have been able to give the wild horses over ten thousand horse years of freedom, but what is really important is this. There are still some of us who care.

Chapter Twenty-Three

From the time the land making up the Black Hills Sanctuary was settled in 1878 to the time I took it over, Indians had not been welcomed there. Perhaps the early white settlers on the land had too many bitter memories of massacres along the trail west. And yet for thousands of years, this land had been one of the great centers of native activity. The flint mines atop the sanctuary, with pits as big as a three-bedroom house, are thought to be twelve to fourteen thousand years old.

To me it was very important that the sanctuary be made available not only to the wild horses but for native ceremonies. Now, every year in June, the Sioux have a sun dance on a ridge overlooking the Cheyenne River. From my little prairie house along the river, I can hear their drums and the sound of chanting. Smoke rises in the air from a host of campfires, and I have a sense that I have done a little something to make peace with history.

My friend Ernest Afraid of Bears appreciates what I am doing here with the wild horses, and his people who run the sun dance make sure when they leave that there is not a soft drink can or a gum wrapper left on the land. Hundreds of

Native Americans attend the dance, camping in a host of tipis rising like white mushrooms upon the Cheyenne plain.

Ernest has a sweat lodge along the river, and invited me to sweat with his people. I knew I wasn't tough enough, but I went. For hours the participants heated rocks in a huge cottonwood fire, and about midnight we went into the sweat lodge, the chief first, then me as the guest of honor, until there were probably fifteen of us crowded into the tiny lodge.

The lodge was made of chokecherry and willow saplings bent into a dome and tied, then covered with hides and canvas. A pipe was passed around the circle, then the hot rocks were brought in with deer antler tongs and placed in a pit in the center, while men chanted prayers in Sioux and poured water on the hot rocks. It was soon so hot that I was ready to pass out. Now and then someone would pass a tin dipper of cool water around the circle, which only enhanced my desire for more. I hugged the ground, trying for cool air to help my burning lungs.

I wanted out, but I didn't want to hurt anyone's feelings or be thought a wimp. I eyed the door desperately, thinking that if I tried to climb out half-naked over the chief, he would go right through the roof. It got hotter and hotter, and there was no sweat left in me. I panicked and, like a berserk badger, dug my way out under the edge of the hut, then leaped into the river. I drank and drank, hoping I wouldn't come down with beaver fever.

No one came out, and had it not been for the chanting, I would have thought them all dead. There was a big pot of stew simmering on the coals of the fire, and I sat there alone, helping myself to bowl after bowl. At last Ernest came out of the lodge as I was helping myself to another bowl. "Dig deeper in the pot," he grinned. "You get the best puppy!"

The history of the land is written in fossils, petrified wood, seashells, and ancient corals from the time when this was a warm, shallow ocean stretching to what is now the Caribbean, some sixty-five million years ago. Signs of early man, in the form of petroglyphs, flint points, and stone implements such as axe heads used in butchering woolly mammoths are present, but even they are recent compared to the brutal canyons carved into the land by the nervous fingers of the wind.

I hunt arrowheads constantly, but my eyes are a long way from the ground. In my sixteen years here I have found only a handful of ancient points, bone needles, a pestle and mortar, a stone axe, an ancient agate lamp, and the ironstone figure of a pregnant woman. I have probably stepped over hundreds more. Some folks have a special knack.

Back in Oregon I knew an old cowboy named Holly Brown. In his youth, Holly was riding a horse and fell into a well, and when folks fished Holly out, he was bent stiff at a ninety-degree angle. Holly did the best he could with his affliction and sold Watkins Products around the country. He could even joke a little about his condition. "I don't know anything about birds in the treetops," he said, "but I bet I can find more arrowheads than most folks, twenty to one."

I relish the presence of live Indians as much as traces of their ancestors. Not long ago an elderly Sioux couple arrived at the sanctuary to see the wild horses. She was ninety-three, and her husband was two years her junior. There was a sparkle to them both, and a love of life that made it fun to be with them. I drove them back into the mountains in my old, experienced pickup truck. To make the woman more comfortable sitting in the middle, I took off my buckskin gloves and placed them over the ends of the seat belt receptacles protruding from the seat.

She had been sitting on the gloves for some minutes, as we bounced over bumps in the road, when she turned to her husband and said with a twinkle, "Henry! This is more fun than I've had in thirty years."

I had never eaten antelope meat, and the couple invited me down to the reservation for supper that night. I was sitting in the parlor reading some historic reservation records, and they were in the kitchen frying some antelope over an old woodstove. As Henry passed behind his wife, he reached out and pinched her bottom. She retaliated by patting him on the front of his trousers. "Oh, my, Henry," she grinned. "That make very fine bone for soup!"

Sometimes things happen that I can't begin to explain. There is a cave overlooking the Cheyenne that was used for ceremonies by ancient people. Petroglyphs adorn the walls, and in one corner are old dried buffalo bones, cracked for their marrow. Just inside the entrance of the cave, the soil is dark from ashes of ancient campfires; in front of the cave is a huge rectangular rock shaped like an altar. The massive rock has been split in two by the roots of an ancient hackberry tree.

There is something powerful about the place that attracts me. Not long ago, seeking a better understanding of the cave's energies, I made a tobacco offering at the entrance, hoping that the spirits of the cave would welcome me. Many years ago, I had a young Blackfoot friend named Robert Butterfly, who eventually died of alcoholism on Seattle's skid row. As I entered the cave, I was feeling very white, and asked Robert Butterfly's spirit to join me.

I had scarcely seated myself inside when a small black skipper butterfly appeared before me and, for some minutes, danced before my face. I thought I heard the guttural murmuring of Indian voices, and saw unexplained facelike shadows moving on the rock walls. When the butterfly left, a group of seven rock wrens flew into the cave and hopped about me, showing no fear. They perched on my legs, my arms, my head, and my shoulders, and peered at me from adjoining rocks. When they departed, a sudden blinding wind filled the cave, choking me with dust, and obliterating all traces of the bird tracks in the sand. Then, as suddenly as the wind had started, it was gone, and out of the starkness of the silence there came the deafening roar of cicadas from the surrounding pines and hackberry trees. The shadows of the Indian faces flickered for a moment on the rock walls, then all was quiet and the cave seemed deserted. But I was filled with peace. It was as though the ancients knew my love for this great sanctuary, and were thanking me for protecting it.

As I left the cave, a haze of old friends sat there on their horses waiting for me, smiling. I might have gone with them at that moment, but I knew that the wild horses still needed me, and that my job here on the sanctuary was a long ways from being finished.